Marc Achtelig

Reihe „Lösungen zur Technischen Dokumentation":
Designing Templates and Formatting Documents

Wie Sie Benutzerhandbücher und Online-Hilfen attraktiv und gut lesbar gestalten, und wie Sie effiziente Formatvorlagen erstellen

Erste Auflage
Zweisprachig: Englisch + Deutsch

indoition

indoition publishing e.K.
Goethestr. 24
90513 Zirndorf bei Nürnberg

Tel.: *+49 (0)911/60046-659*
Fax: *+49 (0)911/60046-863*
E-Mail: *info@indoition.de*
Internet: *www.indoition.de*

Autor: Marc Achtelig
Lektorat: Andrea R. Winter
Korrektorat: Elizabeth Meyer zu Heringdorf, Johannes Schubert
Satz: Marc Achtelig
Titeldesign: Marc Achtelig
Titelbild: alexsl, iStockphoto
Druck: Lightning Source

Marken
Alle in diesem Buch erwähnten, dem Verlag und Autor als Marken oder Warenzeichen
bekannten Begriffe, wurden in Textabschnitten in englischer Sprache durch entsprechende
Großschreibung gekennzeichnet. In Textabschnitten in deutscher Sprache erfolgte analog
ebenfalls Großschreibung. Weder Verlag noch Autor übernehmen jedoch eine Gewähr für
die Richtigkeit und Vollständigkeit dieser Kennzeichnung. Die Verwendung eines Begriffes
in diesem Buch lässt nicht auf das Bestehen oder Nichtbestehen eines markenrechtlichen
Schutzes schließen.

Warnhinweise und Haftungsausschluss
Die Informationen in diesem Buch wurden mit größtmöglicher Sorgfalt erstellt und auf
Richtigkeit und Vollständigkeit geprüft. Weder Verlag noch Autor übernehmen jedoch eine
Gewähr für Richtigkeit und Vollständigkeit gleich welcher Art. Ebenso übernehmen weder
Verlag noch Autor eine Gewähr für Gültigkeit und Anwendbarkeit der enthaltenen Inhalte.
Weder Verlag noch Autor übernehmen eine Haftung oder sonstige Verpflichtung gegenüber
natürlichen oder juristischen Personen im Hinblick auf Schäden oder Verluste jeder Art, die
als Folge aus den in diesem Buch gegeben Informationen entstehen, oder die aus dem
Fehlen von Informationen entstehen. Das Buch bietet keine individuelle Beratung,
insbesondere keine rechtliche Beratung. Stellen Sie vor Auslieferung Ihrer Produkte und vor
Veröffentlichung Ihrer Inhalte sicher, dass Sie alle Standards, Normen, Gesetze und
sonstigen Regelungen befolgen, die für Ihr eigenes Land sowie für alle Länder in die Sie Ihre
Produkte verkaufen und liefern maßgeblich sind. Alle in diesen Standards, Normen,
Gesetzen und sonstigen Regelungen enthaltenen Bestimmungen haben Vorrang gegenüber
den in diesem Buch gegebenen Empfehlungen.

Bibliografische Information der Deutschen Nationalbibliothek
Die Deutsche Nationalbibliothek verzeichnet diese Publikation in der deutschen
Nationalbibliografie; detaillierte bibliografische Daten sind im Internet über
http://dnb.d-nb.de abrufbar.

ISBN 978-3-943860-07-8

Über den Autor

Marc Achtelig ist Diplom-Ingenieur (FH) für Verfahrenstechnik und Wirtschaftsingenieurwesen und seit 1989 im Bereich der Technischen Kommunikation tätig.

Nach mehrjähriger Tätigkeit bei einem der größten deutschen Dokumentations-Dienstleister als Technischer Redakteur, „Information Architect" und Berater gründete er 2004 sein eigenes Consulting- und Dienstleistungsunternehmen.

Marc Achtelig war bereits in den 90er Jahren einer der Pioniere im Bereich Single-Source-Publishing, dem Ansatz, gedruckte Handbücher und Online-Hilfen aus einer gemeinsamen Textquelle zu erzeugen. Zu seinen Referenzen zählen zahlreiche Fachartikel, mehrere Bücher sowie diverse Vorträge, Tutorials und Workshops auf nationalen und internationalen Tagungen.

Für individuelle Beratungen und Schulungen erreichen Sie Marc Achtelig unter *ma@indoition.de*.

Inhalt / Contents

1 How to use this book

Welcome to this book, your companion to designing user manuals and online help systems that:

- please the eye
- are easy to navigate
- communicate their message clearly
- are efficient not only for their readers, but also easy to set up and maintain for their authors

What you will find in this book

Aesthetics isn't the only thing that you should be striving for when designing a template. Usability, readability, and simplicity are just as crucial. Defining paragraph styles and character styles that are efficient to use for the authors who write and maintain a document requires a lot of experience in technical writing. The rules presented in this book are the essence of this experience and can prevent you from making the same mistakes that have cost others lots of time and trouble.

All chapters provide various examples that you can use for inspiration and as starting points for your own designs.

Each topic begins on a new page, so skimming through the book is easy. You don't have to read everything from start to finish. All topics are independent of each other. You don't have to read any particular topic to understand another one.

What you won't find in this book

The book provides clear rules and unambiguous recommendations. No boring theory, no musings, no designers' shoptalk.

However, because each product and corporate style is different, there are often no ready-made, one-size-fits-all solutions. The book shows you what's important, it introduces you to the basic rules, and it can inspire you. However, the book can't make the final decisions for you.

Please consider the given rules as general recommendations, not as laws that must be followed slavishly.

> **ⓘ Important:** The book can't provide individual advice; in particular, it can't provide any legal advice. Before shipping your products and before publishing any content, make sure that you also follow all relevant standards, laws, and other regulations that are applicable for both your own country and for all countries in which you sell your products. All rules that are given in these standards, laws, and other regulations take precedence over the recommendations given in this book.

This book is about design—but its own design isn't perfect

When reading this book, you will find line breaks, page breaks, and other things that aren't perfect.

- **Sometimes it's not possible to avoid unaesthetic things:** For example, if you have a large picture after a page break, you can make the picture smaller so that it fits onto the rest of the previous page, but sometimes you just can't make it small enough without making it illegible.

- **Sometimes it's not economical to avoid unaesthetic things:** For example, we've used a special authoring tool that can generate this book in various printed and online versions from the same source files. One of the downsides of this tool is that it doesn't have any built-in control for widow and orphan lines. We've accepted this limitation because the tool has saved us lots of time that we could then invest into optimizing the contents of the book.

What can you learn from this for your own documents? Don't be pedantic. Spend your resources wisely. If you provide much value to your readers, they will tolerate more imperfections than if you provide little value. We hope that we will provide enough value so that you will forgive us for the imperfections in *this* book.

Have a good time reading and creating your own documents.

2 Designing

You never get a second chance to make a first impression.

Don't underestimate the importance of design. Good design cannot compensate for poor content, but it's one of the key factors of successful technical communication. An attractively designed communication product motivates users to read it, much like an attractively wrapped present motivates its recipient to open it.

Readers stay longer in an attractive-looking document than they stay in an unattractive-looking one. Readers who are motivated and stay in the document longer are more likely to understand the given information. Users who understand the given information are more likely to use your product successfully.

A clear design and well-thought-out templates also motivate and help you, the author, to produce clear, user-friendly content.

When to design

Many people, especially developers, argue that the design of a document should be the final step before you ship your document. Theoretically, this is true; practically, however, we don't recommend this approach for a number of reasons:

- If you haven't prepared appropriate styles for paragraphs, characters, links, and tables, you will have to go through all of your texts again just to apply the formatting. If you set up your templates beforehand, you can assign the appropriate styles right away.

- While writing, you don't see what you get. If you don't see the final result, this makes it more difficult to assume the perspective of the reader and to decide whether your text is easy to comprehend.

- Writing an unformatted text is simply less motivating than writing an attractive-looking one.

Tip:
Usually, the most efficient approach is to design the template beforehand without attempting to make the design pixel perfect at this stage. Instead, aim for about 80% of the quality standard that you want to achieve. Then, start writing and iteratively improve the templates where necessary. When writing is finished, review the design and make final adjustments.

What you need to know

You don't need to be a graphic artist to create a professional-looking, user-friendly template for a user manual or online help file.

- *Layout basics* 15
  Shows you the overall principles that you should follow when designing a template.

- *Setting the type area* 59
  Shows you how to segment the page or screen.

- *Choosing fonts and spacing* 85
  Shows you the basic typographic conventions that you should set up.

- *Recommended screen layouts* 103
  Provides examples of basic screen layouts for your inspiration.

- *Recommended page layouts* 129
  Provides examples of basic page layouts for your inspiration.

- *Recommended paragraph styles* 179
  Shows you which paragraph styles you should typically set up, and which paragraph settings are important.

- *Recommended character styles* 219
  Shows you which character styles you should typically prepare, and which settings are recommended here.

- *Recommended table styles* 159
  Provides some examples of how you can create professional-looking tables.

2.1 Layout basics

When setting up a template, two things are equally important:

- The template must be user-friendly. It must make finding information easy, and it must make reading easy.
- The template must be writer-friendly. It must be easy and efficient to use, and it must automate formatting as much as possible.

Don't trade in one for the other.

Things that make a template user-friendly

A user-friendly template:

- looks appealing, professional, and trustworthy
- makes it easy to skim a document for specific information
- makes it easy to decide what information is important and what information is less important
- makes reading easy and doesn't distract the reader

Things that make a template writer-friendly

A writer-friendly template:

- provides only a manageable number of styles
- uses a well-thought-out convention for style names so that writers can easily remember style names
- provides keyboard shortcuts to efficiently assign formats
- provides styles that minimize the need to manually tweak line breaks and page breaks
- provides styles that can be changed later without having to revise the whole document

> **ⓘ Important:** Don't underestimate the importance of an efficient template. Typically, user assistance needs to be updated frequently—in particular, software user assistance. Each template inefficiency will multiply with the number of updates you make. If your document is translated into foreign languages, inefficiencies will also multiply with the number of translations.

Key principles

The key rules for designing user-friendly and writer-friendly templates are:

- *Design for your audience* 17
- *Proceed top down* 18
- *Use clear and simple design* 19
- *Use color with care* 21
- *Visually support skimming* 24
- *Guide the reader's eye* 25
- *Align texts and objects to a design grid* 30
- *Use the golden ratio* 33
- *Trust your visual judgment* 35
- *Think ahead about printing* 36
- *Avoid amateurish formatting techniques* 37
- *Automate line breaks and page breaks* 41
- *Create styles semantically* 47
- *Create styles hierarchically* 56

Related rules

Setting the type area 59
Choosing fonts and spacing 85

2.1.1 Design for your audience

Don't create a template to win a design award. Don't create a design that designers like; create a design that your audience likes.

Take into consideration:

- Who's your audience?
- Where and under what conditions does your audience read the information?

Examples of requirements resulting from the audience

- Elderly people need a larger font size and larger line spacing than young people.
- Engineers prefer a more frugal design than consumers.
- Unskilled workers need bolder hints than university graduates.

Examples of requirements resulting from where and how the audience reads the information

- Users who read in a loud production hall need a larger font size and more line space than users who are sitting in a quiet office.
- Users who read while sitting in a moving car, bus, train, plane, ship, or other vehicle need a larger font size and more line space than users on solid ground.
- Users who read in sunlight need pages that don't produce a glare.
- Users who read in rooms where lighting conditions are poor need strong contrasts between paper and letters, and between different colors.
- Users who have to carry a printed manual with them may need a document with a small page size that fits in their pockets or in a special compartment attached to the product. For example, the manual of a car should fit into the car's glove compartment.
- Users who view an electronic document on a mobile device need a layout that takes up little screen real estate.

2.1.2 Proceed top down

When designing a template, proceed top down. Building a template is like building a house: Start with the building, then furnish the rooms; only put the pictures on the wall when all the furniture is in place.

Don't hesitate to make scribbles on paper. Often, scribbles on paper inspire creativity much better than designing on screen. This is especially true during the early stages, for example, when developing the general layout of the type area, or when designing the title page.

Typical design steps

Proceed in the following order:

1. If you're designing a template for a printed manual: Decide on which page size you will use.
   If you're designing a template for online help: Decide for which screen size you're going to optimize the layout.
2. Set up the type area. For example, decide on whether to use a singular-column layout or a multi-column layout, whether or not to distinguish right pages and left pages, the width of page margins, the position of headers and footers, and so on.
3. Decide on colors.
4. Decide on fonts, font size, and spacing.
5. Create table styles.
6. Create paragraph styles.
7. Create character styles.

2.1.3 Use clear and simple design

Less is more.

Keep the design as simple and plain as possible. Don't distract the reader.

Don't create a design that's an end in itself. Create a design that helps the reader to retrieve and process the given information.

Avoid variety

- Use only a few colors. Use a different color only if it has a particular purpose. Reserve bold colors for important things.
- Use only a few fonts and font sizes. Use a different font or font size only if it has a particular purpose. Reserve bold, large type for important things and for headings.
- Try to get along with as few styles as possible.
- Use the same styles throughout the whole document.
- Use the same styles also for all other documents that relate to the same product.
- In pictures, always use the same line width.
- Always use the same positions when arranging objects. Align all objects on a common design grid.

Avoid redundancy

Omit every letter, symbol, line, or other object that doesn't actually add any value.

- Objects that convey some important or helpful information *do* add value.
- Objects that help readers to find or process information *do* add value.
- Objects that are just there to fill some empty space *don't* add value.

For example:

- Don't use background images.
- Don't put your company logo on each page.
- Don't mention the author's name on each page.
- Don't put a copyright notice on each page.
- Don't put a revision number or release date on each page.

- Don't put the document title on each page.

  A running header or footer that shows the title of the current section, however, can be helpful because it keeps readers oriented.

Avoid clutter

Don't overload your pages with too much information.

Use white space purposefully to direct the readers' attention, to group things that belong together, and to set reading pauses.

It's no problem if an uncluttered layout makes your document a bit longer. If you provide valuable content, readers accept that they need to scroll or turn pages. However, they don't accept documents that overwhelm or confuse them.

Related rules

Use color with care 21

2.1.4   Use color with care

Use color to *communicate*. Don't use color to *decorate*.

Use few colors. Usually, you don't need more than two or three. Often, the best solution is to use gray as a third "color" so that you only have black, another color, and gray.

Use unobtrusive colors.

Reserve striking colors for things that are especially important, such as warnings.

Don't use color as the only coding. When a text is printed on a black and white printer, color is lost. Also, about up to 8% of men are color blind or partly color blind.

- Make sure that the chosen colors have a different gray scale value.
- Apply a second coding:
  - For text, a second coding can be bold or italic font style.
  - For lines, a second coding can be a different line width or line style, such as dashed or dotted.
  - For areas, a second coding can be a different fill pattern.

Font color

On paper, the bigger the contrast between the text and the background is, the easier it is to read the text. On screen, however, if the contrast is too strong, this may result in slight flickering.

Tip:
In a printed manual, use black font color on white paper. In online help, use dark gray on a white background. A good gray value, for example, is an RGB value of 34,34,34 (or 222222 as a hexadecimal value).

Dark text on a light background is easier to read than light text on a dark background. For this reason, use inverted text sparingly. However, for small phrases, the poorer readability is almost irrelevant. So, an inverted book title on a cover page, or an inverted row title or column title in a table is OK. If used only for short phrases, inverted text can be an excellent way to provide a high contrast without the need to add an extra color.

Use existing colors

Often, a corporate style, a company logo, a product logo, or the product itself already use certain colors. In this case, don't add new colors but use the existing ones. Your document then appears as an integral part of the product family.

Tip:
Often, it's a good idea to pick out the most prominent color from the product logo and to use this color as the only color within your document except for black and gray. For example, you can use the highlighting color for level-1 headings, for your product's name, for bullets in bulleted lists, for numbers in numbered lists, or for callout lines in pictures.

Tip:
If you need a second color, choose a lighter or darker version of the first color, or choose the complementary color of the first color.

Choose harmonious colors

Don't rely on your individual preferences but use a professional tool or table for finding a set of harmonious colors. Some image editors and desktop publishing programs even come with a built-in color tool. If your program doesn't have such a tool, you can find various free color scheme generators on the Internet. Search for "color scheme generator" or "color scheme designer."

Tip:
If your corporate style guide, your company logo, your product logo, or your product itself already include a defined base color, use this color as the primary color and look for colors that blend well with this color.

Related rules

Use clear and simple design 19

2.1.5 Visually support skimming

Provide as much orientation as possible.

Provide clear visual means that help readers to skim your document for the information they need.

Most readers don't read manuals; they scan them quickly until they find an item of interest. But even then they don't start reading. First, they scan the item to decide whether it's worth investing in the time and effort to read the text. Only if they expect to gain some benefit from reading do they actually read.

This applies to printed manuals and online help alike.

Things that make skimming easy

The most important things that you can do to make your documents skimming-friendly are:

- Distinguish headings clearly from the text.
- Likewise, distinguish subheadings clearly from the text. If your authoring tool supports it, consider placing subheadings into a margin column.
- Clearly distinguish different heading levels.
- Choose a clear and simple design with only a few colors and fonts. You can then better highlight important key words and key phrases.
- Provide orientation so that readers can easily identify their current position within the document. In a printed manual, add running headers and footers. In online help, consider providing a breadcrumb trail.
- Create paragraph styles that visually communicate the information type of a paragraph's content. For example, create paragraph styles that clearly set apart warnings, notes, and tips.
- Provide separate paragraph styles for lists and procedures.
- Make page numbers large enough so that all readers can clearly read them. Page numbers aren't dispensable; they're a key navigation tool.

 Related rules

Guide the reader's eye 25

2.1.6 Guide the reader's eye

Direct readers through the pages of your document. Add visual clues that show them which path to take.

If there's a picture or other prominent element that has significant visual weight, this is the place where the reader's eye typically enters the page. If there isn't any outstanding element, most readers start in the upper left corner and look for headings and subheadings.

Once you've set up the starting point, guide the readers with the help of lines and white space, and with the help of the so-called Gestalt principles (Deutsch: *Gestaltgesetze*). In particular:

- Use the *principle of similarity* to point out what works equally.

- Use the *principle of proximity* to visualize what belongs together.

Use all elements consistently. This makes orientation easy once the reader has understood the general principle, and it generates a feeling of familiarity and confidence. Consistent design also looks much more professional than arbitrary design.

Use lines and white space purposefully

- Use lines to guide the readers' attention to a particular object.

- Apply white space to separate objects that don't belong together. Reduce white space to group objects that are related.

- If space is limited, use lines to separate objects that don't belong together and can't be moved apart.

- Use frames to group objects that belong together.

- However, note that each line or frame adds more clutter to the page. If possible, use white space around a group, or add a light common background color to a group instead (place the group on a filled rectangle).

Tip:
Lines often don't need to be complete to achieve the intended effect of guiding the eye. Using a short line instead of a long one can also help to avoid some clutter.

In the following picture, a large amount of white space separates two groups of content. A short line makes the distinction even stronger without adding much clutter.

	Lorem ipsum dolor sit amet, consectetur adipiscing elit. Praesent vel tempor orci. Nunc sed ipsum tincidunt augue feugiat convallis. Duis velit ligula, fringilla vel cursus ac, dapibus non nunc. Sed vitae orci sed urna luctus vestibulum in sit amet nunc. Aenean ut arcu non orci tincidunt placerat. Aenean velit arcu, hendrerit id aliquet id, volutpat in dolor. Sed sed nisi velit, vel rutrum orci. Proin molestie, metus nec dictum consequat, magna ante egestas odio, eget convallis justo magna rhoncus dolor. In imperdiet elit in tortor vestibulum non dapibus lorem vestibulum. Mauris volutpat feugiat sagittis. Donec ut ante eu nibh rutrum ultrices. Vestibulum dapibus hendrerit adipiscing.

Lorem
ipsum
dolor
sit
amet
consectetur
adipiscing
elit
Etiam
vitae
lectus
velit
hendrerit
luctus
tortor
Suspendisse
sagittis
ultrices
tortor
ac
aliquam
In
et
ligula
massa
In
in
nibh
in
augue
blandit
bibendum
at
et
nulla

Vivamus vitae lorem arcu, vitae mattis libero. Curabitur sed magna vel erat ullamcorper blandit eget et ligula. Suspendisse porttitor dolor elit. Fusce dignissim fringilla arcu ut fringilla. Suspendisse nec nunc magna. Ut et risus vel elit molestie vehicula. Vivamus cursus justo lacus, ac consequat lorem. Suspendisse potenti. Duis nec laoreet velit. Aenean gravida felis eros. Mauris nec gravida eros. In nulla tellus, malesuada eu interdum ac, iaculis quis ante. Maecenas aliquet dui nec justo suscipit nec adipiscing nisi elementum.

Pellentesque habitant morbi tristique senectus et netus et malesuada fames ac turpis egestas. Cras vestibulum, purus nec suscipit accumsan, tellus justo bibendum lorem, semper luctus est nibh at lectus. Aliquam erat volutpat. Nunc malesuada facilisis tellus, ut aliquam arcu vestibulum eget. Lorem ipsum dolor sit amet, consectetur adipiscing elit. Aenean id nunc id elit semper mattis eget in eros. Fusce vitae magna ac arcu interdum lacinia dapibus vel ante.

In the next picture, a group is indicated by an incomplete frame.

If you can't position related objects next to each other, use identical colors, shapes, sizes, or fonts to group them. The more things the members of each group have in common, the stronger the visual coherence is.

Find the right visual weight

Put the most visual weight on:

- the objects that readers should notice first
- things that you want to stress

Decrease the visual weight gradually while guiding the readers' eyes from object to object.

As a result of our experience with gravity, we've developed a sense of where objects should "naturally" be. The visual weight of a big object near the top of a page is greater than the visual weight of the same object near the bottom of the page.

As rules of thumb, an object's visual weight increases:

- the closer to the upper left corner of the page that the object is positioned
- the larger the object is
- the more colorful and the darker the object is

In the first picture of the following example, the visual weight of the dark title section with its big letters is definitely too big. It seems to crush the section below. The layout isn't harmonious.

In the second picture, the distribution is much more natural. The objects' sizes, colors, and positions direct the reader's eye systematically as intended from the picture > to the product name > to the document title > to the version number.

✖ No:

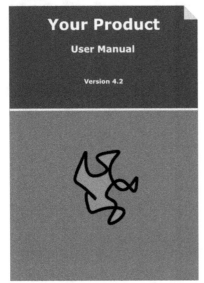

✔ Yes:

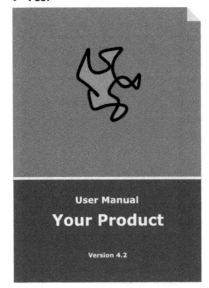

Related rules

Visually support skimming 24

Align texts and objects to a design grid 30

2.1.7   Align texts and objects to a design grid

Don't position text frames and other objects arbitrarily.

Align all objects to a common design grid so that they have common distances from the page or screen margins.

Typical applications of a design grid

Some typical examples of objects that should be aligned to a common design grid are:

- graphical elements and text elements on a title page
- headers and footers that are aligned with the text column
- logos, dates, version numbers, and page numbers
- texts within a picture

For the same reason, always align tables and pictures to the left of the text column instead of centering them. All elements then have a common starting line on the left side of the page, which looks much more consistent and less cluttered than a mixture of left-aligned and centered elements.

Example

In the following example, the lines indicate the common alignments on the different page types of a user manual:

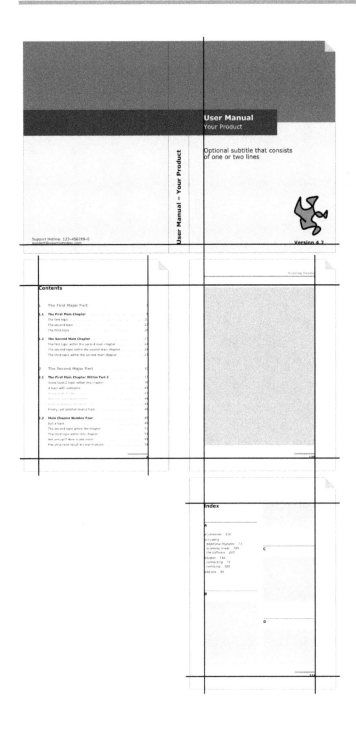

Related rules

Guide the reader's eye 25

2.1.8   Use the golden ratio

When in doubt, base decisions on positions and lengths on the golden ratio. This ratio usually provides a very aesthetic impression because in nature (for example, in plants) many things have this ratio as well.

Two distances or areas are in the golden ratio if the ratio of the total of the quantities to the larger quantity is equal to the ratio of the larger quantity to the smaller quantity. This ratio is always approximately 1.618.

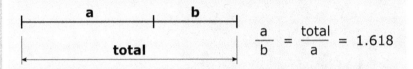

$$\frac{a}{b} = \frac{total}{a} = 1.618$$

With a simple formula, you can calculate the golden ratio as follows:

- If you know the shorter distance (b), multiply it by 1.618 to calculate the longer distance (a).

- If you know the longer distance (a), divide it by 1.618 to calculate the shorter distance (b).

- If you know the total distance (a+b), divide it by 1.618 to calculate the longer distance (a), and divide it by 2.619 to calculate the shorter distance (b).

You don't have to stick to the exact numbers. If the actual dimensions of your objects come close enough to the golden ratio, the human eye won't see the difference.

(DE)

Deutscher Begriff: *Goldener Schnitt*

Typical applications of the golden ratio

You can apply the golden ratio, for example, to determine:

- the position of the title on a title page
- the position of the main object within a picture
- the positions and dimensions of areas on a page
- dimensions of graphical elements in a picture

- paper size (unless you don't have to use a particular standard size; see
  What page size? 61)

Example

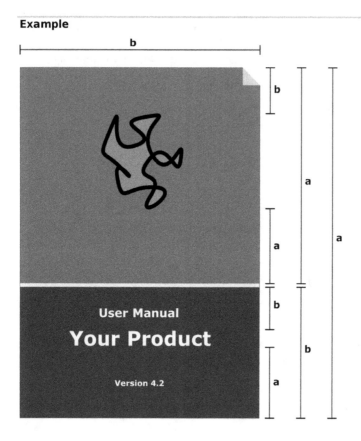

Related rules

▶ *Trust your visual judgment* 35

2.1.9   Trust your visual judgment

Even if something is scaled and aligned precisely, sometimes it looks misplaced or inharmonious to the human eye.

If this happens, trust your visual judgment. Use the setting that looks right rather than the setting that's mathematically correct.

Examples for settings that shouldn't be perfectly symmetrical

When you want to position objects vertically in the middle of an area or frame, the visual middle is slightly above the geometrical middle.

- On pages, make lower page margins slightly larger than upper page margins.

- In tables, make the distance to the lower cell border slightly larger than the distance to the upper cell border.

- Within boxes, position centered text slightly above the geometrical middle.

This text is positioned strictly in the geometrical middle of the box. At first glance, it seems to be placed too low.

This text is positioned slightly above the geometrical middle of the box. This looks symmetrical although it isn't.

This text is positioned clearly above the geometrical middle of the box. Often, this is the most harmonious solution.

Related rules

Use the golden ratio 33

2.1.10   Think ahead about printing

Many documents are shipped only electronically. However, many users print topics that they frequently need, or they print a document to read while commuting to work, or they print a long document because they find reading on paper easier than reading on screen.

Make printing your documents easy.

Make sure that printing consumes as little paper and as little ink or toner as possible.

Test printing early in the design process. Make sure that all texts and images are readable even when printed on a standard black and white office printer. Also make sure that all areas of a page or screen are printed and that no information is cut off.

Tips for designing printer-friendly documents

- With PDF files, use a page size that takes advantage of the full paper size on the users' printers. Usually, Letter format or A4 format is a good choice.

- With online help, consider adding a separate style sheet (CSS) for printing.

- Avoid dark page backgrounds. They may print poorly and consume lots of ink or toner.

- Make sure that the resolution of embedded pictures is high enough to produce clear results. In PDF files, disable downsampling of pictures.

- Make sure that no information gets lost when printing in black and white. In pictures and for character styles, use colors that have different gray scale values. When using color to distinguish character styles, use a second coding. For example, don't create a character style for interface elements that's just blue, but make it blue + bold or blue + italic. When users print the document on a black and white printer, at least the second coding (bold or italic) remains distinguishable.

- If an online document uses expandable sections (toggles), make sure that the contents of these sections are printed along with the document.

- Consider single source publishing. With the help of appropriate authoring tools, you can generate a full-featured printed manual (including a table of contents, page numbers, and an index) from the same source as online help.

Related rules

Right and left pages? 63

2.1.11   Avoid amateurish formatting techniques

Don't use empty paragraphs.

Don't use multiple space characters.

Don't use tabs.

Don't format your text manually. Only use the styles that are configured in your document template.

Problems with empty paragraphs

Don't use empty paragraphs to adjust the space above or the space below a paragraph. Use the appropriate paragraph settings instead.

Also, don't use empty paragraphs to create page breaks. If you later add or delete one or more lines from your text, everything will then move to a wrong position. Instead, always use proper manual page breaks; even better, when possible, completely automate page breaks (see *Automate line breaks and page breaks* 41).

All sorts of empty paragraphs seriously interfere with automatic page breaks. Look at the following example to see what may happen:

✖ No:

This is the preceding paragraph. Let's call it paragraph A.¶
¶
¶
¶

This is the heading
(page break)

¶
And here is the text that follows the heading. Initially, everything
works well. The trouble starts when you later change the text.
Imagine adding a few extra paragraphs above paragraph A so
that paragraph A moves to the bottom of the page. As a result,
the heading won't start properly at the top of the page.¶

This is the preceding paragraph. Let's call it paragraph A.¶
(page break)

This is the heading
¶
And here is the text that follows the heading. Initially, everything
works well. The trouble starts when you later change the text.
Imagine adding a few extra paragraphs above paragraph A so
that paragraph A moves to the bottom of the page. As a result,
the heading won't start properly at the top of the page.¶

✔ Yes:

This is the preceding paragraph.¶
space above paragraph

This is a proper heading¶
space below paragraph

The heading has proper settings for the space above and for the
space below the paragraph. So you don't have to add any empty
paragraphs. If the heading moves to the top of a page, most
authoring tools will remove the space above the paragraph automatically so that the heading starts correctly on top of the page.¶

This is the preceding paragraph.¶
(page break)

This is a proper heading¶

The heading has proper settings for the space above and for the
space below the paragraph. So you don't have to add any empty
paragraphs. If the heading moves to the top of a page, most
authoring tools will remove the space above the paragraph automatically so that the heading starts correctly on top of the page.¶

Problems with multiple space characters

Don't use space characters to indent any text. Always configure proper indentation settings as part of the paragraph properties. If you use space characters for indenting text, as soon as you change a single word all the lines following the word may need to be edited manually.

In online documents that use a variable column width, indents with space characters don't work at all.

The following example shows how even a small change can break up your layout:

✖ No:

This·paragraph·isn't·indented.¶

···This·paragraph·is·indented·with·space·characters.¶

···This·paragraph·is·indented,·too.·But·it's·too·long. Now the trouble starts.¶

···You·could add·manual·line·breaks·and·even·more·space ←
···characters,·but·this·isn't·a·good·solution.·It·requires extra·work,←
···and·if·you·change·the·text,·everything may turn into a terrible ←
···mess.¶

···You·could·add·manual·line·breaks·and·even·more·space ←
···characters,·but·this·isn't·an·acceptable·solution.·It·requires·extra
·work, ⟵
···and·everything·turns ⟵
···into·a·mess.¶

✔ Yes:

The·proper·way·is·to·define·a·paragraph·style·that has·a·left·indent.¶

left indent

The only time when using space characters to format your text makes sense is when you're quoting programming source code. When you copy and paste source code into your document, the source code is usually already indented with the help of space characters or tabs. Also, when formatting source code, you typically use a monospaced font, which has letters and characters that each occupy the same amount of horizontal space. This makes formatting with space characters easier.

Problems with tabs

Avoid using the tab key. Tabs may get you into trouble if you want to produce an online version of your document because tabs don't have any equivalent in HTML. Tabs then need to be converted into tables, space characters, or indents. This rarely works well.

Instead of using tabs, set up proper paragraph indentation, or use a borderless table. Even if you don't need to produce an online version of your document today, you may want to do so in the future.

Problems with customized styles

Don't customize paragraph settings and character settings for individual paragraphs. Always use the paragraph styles and character styles as they're defined in your document template.

If you overwrite a paragraph style for an individual paragraph or if you overwrite a character style for an individual word, this will get you into trouble when you need to change the design of your document later (which may seem unlikely today but often happens even if you don't anticipate it).

If you have exclusively used template styles and want to change the design, all you need to do is to change the style definitions. If you've applied manual formatting, however, you will need to update all manual formattings manually. If you have several documents—maybe even in different languages—this can result in many hours of extra work. So think ahead and apply formats with self-discipline.

Related rules

Automate line breaks and page breaks 41

Recommended paragraph styles 179

Recommended character styles 219

2.1.12 Automate line breaks and page breaks

If you exclusively create online content, the only thing that you need to do is to use nonbreaking spaces where you want to prevent lines from breaking.

If you create printed manuals or PDF files, you also need to take care of hyphenation and page breaks. This can be very time-consuming if you don't automate your templates appropriately.

Be aware that every change that you make to your document may affect line breaks and page breaks. If you don't automate line breaks and page breaks the best that you possibly can, you'll have to revise them over and over again each time you update your document. Don't underestimate the frequency of document updates, especially in software user assistance.

If your line breaks and page breaks need manual tweaking, optimize them only as the final step after you have completely finished writing and editing.

Take a particularly close look at tables. Text within narrow table columns often needs some additional manual line breaks.

Where there should be line breaks

Follow these basic rules as closely as possible:

- In general, insert line breaks after punctuation marks instead of in the middle of a sentence.
- If a heading must have two lines, make the first line longer than the second line.
- Avoid line breaks that disrupt ideas that belong together.
- Don't insert a line break between the parts of a product name and between the parts of a company name.
- Don't insert a line break between a product name and the version number.
- Don't insert a line break between Internet addresses and email addresses.
- Don't insert a line break between numbers and units of measure.
- Avoid line breaks within function names, parameter names, commands, and quoted source code. Also turn off hyphenation here because readers might assume that the hyphen is part of the name.
- Avoid line breaks within cross-references.

✖ No:

> Things that belong ←
> together shouldn't be ←
> separated by a line ←
> break.¶

✔ Yes:

> Things that belong together ←
> shouldn't be separated ←
> by a line break.¶

Automating line breaks

To prevent unintentional line breaks, use all the automation options that your authoring tool provides.

- Add nonbreaking space characters instead of normal space characters (possible with most authoring tools).

- Define and use paragraph styles and character styles that don't allow hyphenation and line breaks (only possible in more advanced authoring tools). For example, create a special character style for your product name, and turn off hyphenation for this style.

Where there should be page breaks

Follow these basic rules as closely as possible:

- Try to keep related material on one page. If you can't avoid a page break, break pages so that readers can anticipate that an idea or topic continues on the next page.

- If you have a layout that has left pages and right pages, handle page breaks on right pages with special care. (Page breaks on left pages are less critical because readers can directly see the next page.)

- Don't separate headings from the text that follows.

- Avoid leaving major headings close to the bottom of the page.

- Avoid separating warnings, notes, and tips from the material that they concern.

- Never allow a page break within a warning.

- When possible, don't leave a single line of text at either the bottom of a page (a so-called "orphan") or on the top of a page (a so-called "widow"). If your authoring tool allows, set orphan control and widow control to a value of *2* for all paragraph styles.

- Don't leave a single list item or a single step of a procedure standing at the bottom of a page or at the top of a page.

- Don't separate an introductory phrase from the list, procedure, table, or figure that it introduces.

- Avoid breaking tables. If the table is too long, take care that at least two rows of the table are at the bottom and at the top of each page.

- If it isn't obvious what information each column of a table contains, repeat the column headings on the new page.

- If a table has a table title, keep the table title on the same page as the table. If you have a page break within the table, repeat the table title on the new page, followed the by the text "*(continued)*" in italic font style.

- If a figure has a figure title, keep the figure title on the same page as the figure.

- In an alphabetical index, if a main entry is followed by subentries, don't leave the main entry alone at the bottom of a column or page. If you must break up a list of subentries, repeat the main entry followed by the text "*(continued)*" in italic font style.

 (DE)

Im Deutschen ist für „*(continued)*" der Zusatz „*(Fortsetzung)*" üblich.

Automating page breaks

To automate page breaks, use all the advanced options that your authoring tool provides:

- Define paragraph style properties that automatically add a page break before the paragraph so that each paragraph of this style starts on a new page. You can use this for headings.

- Define paragraph style properties that always keep particular paragraphs together with the next paragraph on the same page. You can use this, for example, for phrases that introduce lists and procedures.

- Define paragraph style properties that prevent a page break from occurring within a paragraph. You can use this, for example, for warnings.

- Set up the number of widow lines and orphan lines that may be left at the top or bottom of a page, immediately before or after an automatic page break.

■ (DE)

Widow und *Orphan* haben im Deutschen keine wörtliche Entsprechung.

- Der analoge deutsche Begriff zu *Widow* (wörtlich „Witwe") lautet *Hurenkind* (letzte Zeile eines Absatzes erscheint als erste Zeile auf einer neuen Seite oder in einer neuen Spalte).

- Der analoge deutsche Begriff zu *Orphan* (wörtlich „Waisenkind") lautet *Schusterjunge* (erste Zeile eines Absatzes erscheint als letzte Zeile auf einer Seite).

Adding automatic page breaks before as many headings as possible can save you a lot of time.

- All chapters that are shorter than one page don't need any internal page breaks. So, there's nothing to worry about, nothing to check, and nothing to update.

- Changes within one chapter don't affect any of the following chapters because each of these chapters starts on a new page—just as before.

- Readers can spot chapters that begin on a new page more easily.

The downside of beginning a new page for each chapter is that it can make your document longer, so you need more paper. The shorter your average topic is, the stronger this effect will be.

For these reasons, good rules of thumb are:

- If your document is shipped electronically and you don't expect most readers to print it, start *all* headings on a new page.

- If your document is printed, start major headings on a new page (for example, levels 1 and 2), but start minor headings anywhere on a page. If most topics are shorter than half a page, only start heading level 1 on a new page.

Things that can't be fully automated

Usually, line breaks can be entirely automated. Page breaks, however, can only be automated to a degree of about 80% to 95%.

A need for a manual adjustment mostly results from a large image. When an image doesn't fit onto the rest of a page, the image is automatically shifted to the next page, leaving a lot of unused white space.

In cases like this, you have to decide whether you accept the unused, odd-looking white space or whether you resize the image to a smaller size so that it fits onto the previous page.

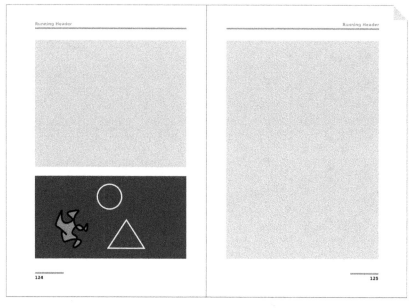

The advantage of resizing the image is that the visual appearance of your document improves.

Disadvantages of resizing the image are:

- Resizing needs extra work.

- The new image size may be inconsistent with the size of other images. For example, if you've used text in the image, the font size will be smaller than in other images.

- Your optimization may not be permanent. If you add, change, or delete some content before the image in the future, the image will move to a different position. You may then want to resize the image back to its original size. In addition, other images may now trigger poor page breaks, and the manual tweaking process starts all over again.

Related rules

Avoid amateurish formatting techniques 37

Hyphenation—yes or no? 73

Recommended paragraph styles 179

Recommended character styles 219

2.1.13   Create styles semantically

Use style names that describe the meaning of styles, not their visual appearance. For example, call a style "Emphasis" rather than "Arial 10 Point Bold." Semantic styles have some major advantages:

- The style name tells you when to use the style meaningfully.

- You can change the appearance of the style at any time without any negative side effects.

- Semantic styles are also a key prerequisite for structured authoring in XML and DITA. If you use semantic styles, your documents are well-prepared in case you want to shift to structured authoring someday.

To make your template writer-friendly, set up a well-thought-out naming convention for style names so that writers can remember them easily.

If your authoring tool supports it, define which paragraph style will be applied to the next paragraph when you press Enter.

Set up a keyboard shortcut for each style so that you can assign styles easily.

Tip:
If your authoring tool doesn't support keyboard shortcuts, use a third-party automation tool, such as the open source automation tool AutoHotkey.

Tips for naming conventions

- Use short style names. This makes it easy to spot a style in the style catalog of your authoring tool, on the status bar, and in other places.

- Create style names so that related styles are listed next to each other in the style catalog. You can achieve this by beginning the names of related styles with the same letters.

Tips for keyboard shortcut conventions

- Use keys that you can easily remember.

- Use the same scheme for all paragraph styles and another scheme for all character styles. For example, use [Alt]+[Shift]+letter for all paragraph styles, and use [Ctrl]+[Shift]+letter for all character styles.

Example scheme for paragraph styles

As a starting point, you can use or adapt the following scheme for your paragraph styles.

Note:
If your authoring tool doesn't allow you to define the suggested keyboard shortcuts, or if your authoring tool already uses these shortcuts, feel free to define other key combinations. The important thing is that you find a consistent scheme that you can easily remember.

Purpose	Suggested style names	Suggested keyboard shortcuts
Headings:		
heading level 1	h1	[Alt]+[Shift]+[1]
heading level 2	h2	[Alt]+[Shift]+[2]
heading level 3	h3	[Alt]+[Shift]+[3]
For details, see: *Heading paragraph styles* 187.		
Subheadings:		
subheading that introduces a static section (always visible)	ss_subhead_static	[Alt]+[Shift]+[u]
subheading that introduces an expandable section (toggle)	se_subhead_ expandable	[Alt]+[Shift]+[e]
For details, see: *Subheading paragraph styles* 194.		
Body text:		
standard paragraph	bo_body	[Alt]+[Shift]+[o]
For details, see: *Body text paragraph style* 199.		
Procedures:		
paragraph that introduces a procedure	pi_procedure_intro	[Alt]+[Shift]+[p]
step of a procedure (with number)	ps_procedure_step	[Alt]+[Shift]+[s]
indented first-level paragraph in a procedure (without number)	pp_procedure_plain	–
For details, see: *Procedure paragraph styles* 201.		

Lists:

paragraph that introduces a list	li_list_intro	[Alt]+[Shift]+[**l**]
first-level item of a list (with bullet)	l1_list_level1_**b**ullet	[Alt]+[Shift]+[**b**]
indented first-level paragraph in a list (without bullet)	l1p_list_level1_plain	–
second-level item of a list (with bullet)	l2_list_level2_bullet	–
indented second-level paragraph in a list (without bullet)	l2p_list_level2_plain	–

For details, see:
List paragraph styles 207

Annotations: **Tips**	at_**t**ip	[Alt]+[Shift]+[**t**]

For details, see:
Note and warning paragraph styles 213

Annotations: **Notes**		
standard note	an_**n**ote	[Alt]+[Shift]+[**n**]
important note	ai_**i**mportant	[Alt]+[Shift]+[**i**]

For details, see:
Note and warning paragraph styles 213

Annotations: **Warnings**		
caution	ac_**c**aution	[Alt]+[Shift]+[**c**]
warning	aw_**w**arning	[Alt]+[Shift]+[**w**]
danger	ad_**d**anger	[Alt]+[Shift]+[**d**]

For details, see:
Note and warning paragraph styles 213

(DE)

Wenn Sie lieber mit deutschen Bezeichnungen arbeiten, können Sie z. B. die folgenden Formatnamen und Tastenkürzel verwenden:

Zweck	Vorschlag für Formatnamen	Vorschlag für Tastenkürzel
Überschriften:		
Überschrift Ebene 1	ue**1**	[Alt]+[Shift]+[**1**]
Überschrift Ebene 2	ue**2**	[Alt]+[Shift]+[**2**]
Überschrift Ebene 3	ue**3**	[Alt]+[Shift]+[**3**]
Details siehe: *Heading paragraph styles* 187.		
Zwischenüberschriften:		
Zwischenüberschrift, die einen statischen Abschnitt einleitet (immer sichtbar)	zs_**z**ue_statisch	[Alt]+[Shift]+[**z**]
Zwischenüberschrift, die einen expandierbaren Abschnitt einleitet (Toggle)	ze_zue_**e**xpandierbar	[Alt]+[Shift]+[**e**]
Details siehe: *Subheading paragraph styles* 194.		
Grundtext:		
Standardabsatz	tx_**t**extkoerper	[Alt]+[Shift]+[**x**]
Details siehe: *Body text paragraph style* 199.		
Handlungsanweisungen:		
Absatz, der eine schrittweise Handlungsanweisung einleitet	si_**s**chritt_intro	[Alt]+[Shift]+[**s**]
Schritt einer schrittweisen Handlungsanweisung (mit Nummer)	sn_schritt_**n**ummer	[Alt]+[Shift]+[**n**]
Eingerückter Fortsetzungsabsatz erster Ebene (ohne Nummer)	sf_schritt_folge	–
Details siehe: *Procedure paragraph styles* 201.		

Listen:		
Absatz, der eine Liste einleitet	li_liste_intro	[Alt]+[Shift]+[l]
Listenpunkt erster Ebene einer Liste (mit Aufzählungszeichen)	l1_liste_ebene1_az	[Alt]+[Shift]+[a]
Eingerückter Folgeabsatz erster Ebene (ohne Aufzählungszeichen)	l1f_liste_ebene1_folge	–
Listenpunkt zweiter Ebene einer Liste (mit Aufzählungszeichen)	l2_liste_ebene2_az	–
Eingerückter Folgeabsatz zweiter Ebene (ohne Aufzählungszeichen)	l2f_liste_ebene2_folge	–
Details siehe: *List paragraph styles* 207		
Anmerkungen: **Tipps**	at_tipp	[Alt]+[Shift]+[t]
Details siehe: *Note and warning paragraph styles* 213		
Anmerkungen: **Hinweise**		
Standardhinweis	ah_hinweis	[Alt]+[Shift]+[h]
Wichtiger Hinweis	aw_wichtig	[Alt]+[Shift]+[w]
Details siehe: *Note and warning paragraph styles* 213		
Anmerkungen: **Warnhinweise**		
Vorsicht	av_vorsicht	[Alt]+[Shift]+[v]
Warnung	aw_warnung	[Alt]+[Shift]+[w]
Gefahr	ag_gefahr	[Alt]+[Shift]+[g]
Details siehe: *Note and warning paragraph styles* 213		

Example scheme for character styles

As a starting point, you can use or adapt the following scheme for your character styles.

Note:
If your authoring tool doesn't allow you to define the suggested keyboard shortcuts, or if your authoring tool already uses these shortcuts, feel free to define other key combinations. The important thing is that you find a consistent scheme that you can easily remember.

Purpose	Suggested style name	Suggested keyboard shortcut
To highlight the name of the documented product and to disable automatic hyphenation for the product name (optional). For details, see: *Product name character style* 223	cp_product	[Ctrl]+[Shift]+[p]
To highlight all user interface and interaction elements, such as window titles, menu items, buttons, keys, levers, and so on. For details, see: *Element character style* 225	ce_element	[Ctrl]+[Shift]+[e]
To mark parameters, such as parameters of function calls, formulas, and so on. For details, see: *Parameter character style* 226	ca_parameter	[Ctrl]+[Shift]+[a]
To mark product-specific technical terms. For details, see: *Term character style* 228	ct_term	[Ctrl]+[Shift]+[t]
To emphasize words and expressions if this is necessary to avoid confusion. For details, see: *Emphasis character style* 230	cm_emphasis	[Ctrl]+[Shift]+[m]

To highlight important keywords that help readers to skim the text (optional). For details, see: *Strong character style* 232	cs_strong	[Ctrl]+[Shift]+[s]
To mark input that users must type. For details, see: *User input character style* 234	ci_input	[Ctrl]+[Shift]+[i]
To mark quotes of source code and configuration files. For details, see: *Code character style* 235	cc_code	[Ctrl]+[Shift]+[c]
To mark links / cross-references to other topics within the document as well as links to external web sites and documents. For details, see: *Link character styles* 237	cl_link	[Ctrl]+[Shift]+[l]

■■ (DE)

Wenn Sie lieber mit deutschen Bezeichnungen arbeiten, können Sie z. B. die folgenden Formatnamen und Tastenkürzel verwenden:

Zweck	Vorschlag für Formatnamen	Vorschlag für Tastenkürzel
Um den Namen des beschriebenen Produkts hervorzuheben sowie um die automatische Silbentrennung für den Produktnamen zu deaktivieren (optional) Details siehe: *Product name character style* 223	zp_produkt	[Ctrl]+[Shift]+[p]
Um alle Schnittstellen- und Bedienelemente zu kennzeichnen, wie Fenstertitel, Menüpunkte, Schaltflächen, Knöpfe, Bedienhebel, usw.	ze_element	[Ctrl]+[Shift]+[e]

Details siehe: *Element character style* 225		
Um Parameter zu kennzeichnen, wie z. B. Parameter für Funktionsaufrufe, Formeln, usw. Details siehe: *Parameter character style* 226	za_**pa**rameter	[Ctrl]+[Shift]+[**a**]
Um produktspezifische Fachbegriffe zu kennzeichnen Details siehe: *Term character style* 228	zi_begr**i**ff	[Ctrl]+[Shift]+[**i**]
Um Wörter und Ausdrücke hervorzuheben, wenn diese Hervorhebung Missverständnissen vorbeugen kann Details siehe: *Emphasis character style* 230	zb_**b**etonung	[Ctrl]+[Shift]+[**b**]
Um wichtige Schlüsselwörter hervorzuheben, die den Lesern helfen, den Text schnell zu überfliegen (optional) Details siehe: *Strong character style* 232	zh_**h**ervorhebung	[Ctrl]+[Shift]+[**h**]
Zur Kennzeichnung von Benutzereingaben über die Tastatur Details siehe: *User input character style* 234	zg_ein**g**abe	[Ctrl]+[Shift]+[**g**]
Zur Kennzeichnung von Quellcode-Zitaten und Zitaten aus Konfigurationsdateien Details siehe: *Code character style* 235	zc_**c**ode	[Ctrl]+[Shift]+[**c**]

Zur Kennzeichnung von Links / Querverweisen zu anderen Themen innerhalb desselben Dokuments, zu externen Webseiten oder externen Dokumenten	zl_link	[Ctrl]+[Shift]+[l]
Details siehe: *Link character styles* 237		

Related rules

Recommended paragraph styles 179

Recommended character styles 219

2.1.14   Create styles hierarchically

If your authoring tool supports it, don't set up each style independently, but create a hierarchy of styles.

When styles are organized hierarchically, you can later change settings centrally in one place. All child styles inherit the changed setting automatically. For example, if you want to change the font, you only need to change the font of the root element. All other styles that don't use another special font then automatically inherit the new font as well. Thus, instead of having to adapt *all* styles manually, you only need to change *one* style.

Example of hierarchically organized styles

The basic styles (see *Create styles semantically* 47) could be organized as outlined below. If you want to change, for example, the color of all headings, you only need to change the color of the parent style **Headings**. You don't have to change the styles **h1**, **h2**, and **h3** individually.

Custom_Paragraph_Styles
- **Headings**
  - h1
  - h2
  - h3
- **Subheadings**
  - ss_subhead_static
  - se_subhead_expandable
- b0_body
- **Procedures**
  - pi_procedure_intro
  - ps_procedure_step
  - pp_procedure_plain
- **Lists**
  - li_list_intro
  - l1_list_level1_bullet
  - l1p_list_level1_plain
  - l2_list_level2_bullet
  - l2p_list_level2_plain
- **Annotations**
  - at_tip
  - **Notes**
    - an_note
    - ai_important
  - **Warnings**
    - ac_caution
    - aw_warning
    - ad_danger

Custom_Character_Styles
- cp_product
- ce_element
- ca_parameter
- ct_term
- cm_emphasis
- cs_strong
- ci_input
- cc_code
- cl_link

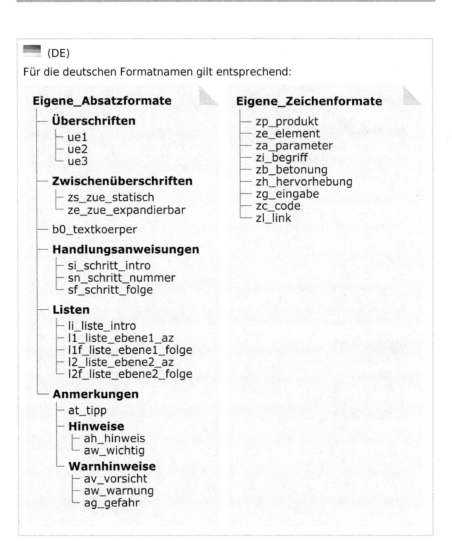

(DE)

Für die deutschen Formatnamen gilt entsprechend:

Eigene_Absatzformate

— **Überschriften**
- ue1
- ue2
- ue3

— **Zwischenüberschriften**
- zs_zue_statisch
- ze_zue_expandierbar

— b0_textkoerper

— **Handlungsanweisungen**
- si_schritt_intro
- sn_schritt_nummer
- sf_schritt_folge

— **Listen**
- li_liste_intro
- l1_liste_ebene1_az
- l1f_liste_ebene1_folge
- l2_liste_ebene2_az
- l2f_liste_ebene2_folge

— **Anmerkungen**
- at_tipp
- **Hinweise**
  - ah_hinweis
  - aw_wichtig
- **Warnhinweise**
  - av_vorsicht
  - aw_warnung
  - ag_gefahr

Eigene_Zeichenformate

- zp_produkt
- ze_element
- za_parameter
- zi_begriff
- zb_betonung
- zh_hervorhebung
- zg_eingabe
- zc_code
- zl_link

Related rules

Recommended paragraph styles 179

Recommended character styles 219

2.2 Setting the type area

The type area is the "stage" for presenting your information. Its size and shape have a great impact on how users perceive what you present.

 (DE)

Der deutsche Begriff für *Type Area* ist *Satzspiegel*.

Large type area vs. small type area

A large type area enables you to present a lot of information simultaneously. Use a large type area if you need to provide:

- overviews
- long procedures
- lots of reference information
- large detailed pictures
- long or wide tables

A small type area enables you to focus on specific information. Use a small type area if:

- your audience isn't able to process much information at a time
- you want to create the impression that the use of your product is very simple
- you must adapt to a small paper size or small screen size for other reasons

Decisions involved

Setting the type area involves a number of decisions that influence each other. It's good practice to proceed in the following order:

1. Determine the page size
   (see *What page size?* 61).

2. Decide whether to provide left pages and right pages or only one template for all pages
   (see *Right and left pages?* 63).

3. Decide whether to use multiple text columns on one page, and whether to provide a margin column
   (see *One or multiple text columns?* 65).

4. Decide how to align your text, and whether to hyphenate
   (see *Left-aligned text or justified text?* 71 and *Hyphenation yes or no?* 73).

5. Roughly determine the size of the page margins
   (see *What page margins?* 75).

   The size of the required margins also depends on the chosen font, so you might have to fine-tune it in a second pass after choosing the font (see *Which font size and font spacing?* 93 and *What line length?* 95).

6. Design headers and footers
   (see *Where to position headers, footers, and page numbers?* 80).

Tip:
When in doubt, make a blur test: Look at a design and squint, or blur a screen capture or picture with an image editor on your computer. Don't look at the details of your design but at the appearance as a whole. Often, a simple paper prototype also helps. Observe how the different areas on your design attract your attention. Choose the one that comes closest to what you've intended.

Related rules

Recommended page layouts 129

Recommended screen layouts 103

2.2.1 What page size?

If your document will be shipped electronically, use A4 format or Letter format, depending on which format is more common in the countries where you sell your product. If you leave page margins of at least 1.5 cm on all sides, both formats usually also print well on the paper of the other format.

Alternatively, you can use a custom format of 21 cm x 27.94 cm, which combines the minimum dimensions of A4 format and Letter format.

If your document will be shipped on paper:

- If there's any corporate style guide that you're obliged to follow, use the page size that this style guide suggests.

- Check with product management and sales whether the manual needs to fit into any given package, case, or compartment.

- Check with your printers' shop whether they have preferred sizes for which printing costs are especially low.

- Within the remaining scope and options, choose a page size that suits the requirements of your audience as best as possible (see also *Design for your audience* 17). Do the users need a small page size so that the manual fits into a pocket? Or do users need a large page size with lots of information on one page—for example, in a service manual, where turning pages with dirty fingers is very inconvenient?

Tip:
When in doubt, use a larger rather than a smaller page size, especially if your document contains many detailed pictures and many tables.

Usually, use portrait orientation. Use landscape orientation only if you need to show a lot of information side by side.

Standard formats

The most common standard formats are:

- A4: 21 cm x 29.7 cm
- Letter: 21.59 cm x 27.94 cm
- A5: 14.8 cm x 21 cm
- B5: 18.2 cm x 25.7 cm
- B6: 12.80 cm x 18.20 cm
- Pocket: 10.5 cm x 21 cm

Custom formats

If you don't want to use one of the standard formats, decide on either the width *or* the height that you want, and then use the golden ratio to determine the second dimension (see *Use the golden ratio* 33).

Foldout pages

If a small page size is sufficient for most parts of your document but you have a few pages where a larger page size would be better, having one or more foldout pages can be a user-friendly option. Information that you place on the outside of the foldout page remains visible even after turning the next page.

Foldout pages are especially helpful for information that users frequently need in parallel when reading other sections of the manual, such as:

- pictures or drawings of the product that show the names and locations of parts or controls
- tables with technical data

Tip:
Depending on how much information you need to include, you can fold a foldout page multiple times. When the page is unfolded, it can be several times as wide as the body of the manual. Sometimes, brief instructions can even be completely organized as one single foldout instead of a traditionally bound manual.

Related rules

What page margins? 75

2.2.2   Right and left pages?

When you create printed manuals or PDF files, most authoring tools let you create either one common template master page for all pages, or separate master pages for right pages and left pages.

- If your document will be printed double-sided and shipped in bound form, create both right pages and left pages.

- If your document will be shipped electronically, create only one master page for all pages. Design this page as if it were a right page.

Examples

A layout that has right pages and left pages typically has headers and footers aligned to the outer side:

A single-sided layout has only right pages:

Handling empty pages

Some authoring tools let you configure how they handle empty pages.

- If you don't distinguish between right pages and left pages, there won't be any empty pages.

- Empty pages can occur if you distinguish between right pages and left pages and tell one or more paragraph styles to start only on a right page. Usually, empty pages are left entirely empty. Optionally, you can continue the page number on the empty pages.

- If you intentionally provide empty pages for readers to take notes, clearly mark these pages as notes pages.

2.2.3   One or multiple text columns?

In most cases, having a single text column is perfectly OK.

Layouts that feature multiple columns can sometimes be more user-friendly but often result in some extra work for the author.

Online content

In online content, always use a single text column for the content. If you have multiple text columns, users may need to scroll up and down multiple times, which is unacceptable.

Use a second column only for navigation (the table of contents, the index, search, related topic links, and so on).

If you need to place particular information side by side, use a table.

Printed content

In printed content, base your decision on your authoring tool. Not all tools provide adequate support for multi-column layouts. If your authoring tool doesn't support multi-column layouts, it's usually OK to stick with a single-column layout. Don't attempt to push your tool beyond its limits by applying dubious workarounds. This *will* get you into trouble.

If your authoring tool does support multiple text columns, you have several options:

Single column

This is the standard layout. If you don't have much experience with template design, it can save you lots of time and trouble. It's also the most economical version if your document will consist mainly of text and if you need a lot of space for large pictures and tables.

The disadvantage of a single-column layout is that it often produces very long lines of text, which may hurt readability.

Margin column plus single column

In this scenario, the width of the margin column is about 25% to 33% of the page width. You can add extra navigational information or extra comments in the margin column.

This layout supports selective reading (skimming) and is, therefore, especially adequate if you have a lot of reference information and want to use the margin column as a navigation tool.

The disadvantage is that this layout needs a lot of space, so it makes your document longer in terms of the number of pages.

Depending on how you use the margin column, this layout may also be incompatible with a single-source-publishing approach. It can be difficult to use the same text source to generate online content in only a single-column layout.

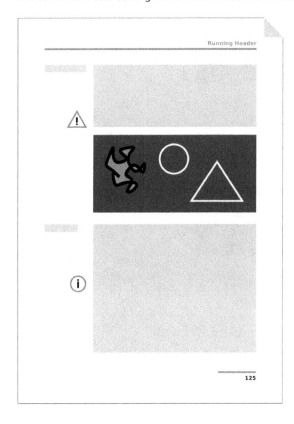

Two or more unsynchronized columns ("Lift")

If you need to include a lot of pictures and have only short texts, this layout gets more content onto one page than a single-column layout. It doesn't waste any space to the right and to the left of the pictures.

If the bulk of your content, however, consists of text, this layout will result in a very large number of line breaks, leaving only a few words in each line. This can make it hard for users to read a longer section of text. In addition, you can't include any large pictures and tables. Their maximum size is always restricted by the column width. This can cause additional readability problems.

Two synchronized columns ("Twin")

Here, you have one column exclusively for pictures and one column exclusively for text. You can always position the text right next to the corresponding picture. This can be a good solution, particularly to illustrate the steps of procedures.

The disadvantage of this layout is that it only works if you always have about the same amount of pictures and text. If you don't, the "twin" layout takes up a lot of space and increases the number of pages, bloating the document.

How much space should you add between columns?

If column spacing is too small, readers sometimes accidentally continue reading in the wrong column.

If column spacing is too large, the text loses its coherence. What's in the next column then doesn't seem to belong together with the text in the previous column.

If column spacing is too large, the text columns may also become too narrow so that there are too many line breaks.

As rules of thumb, make the space between two columns:

- approximately the same size as an empty line plus 2 points
- approximately the same size as the string "mmi" written in the used font

To separate the columns more clearly, you can optionally add a fine line between them.

- Be aware of the fact that the line adds some extra clutter to the page.
- To reduce the visual impact, don't make the line black but use a light shade of gray.

2.2.4 Left-aligned text or justified text?

In all sorts of user assistance, use left-aligned text.

Avoid justified text for the following reasons:

- In good user assistance, most paragraphs are short. Therefore, other than with a novel, for example, the more uniform visual appearance of justified text is almost unnoticeable.

- For the reader, when all lines end at the same position, it makes it more difficult to move the eyes to the next line without landing in the wrong place.

- When lines are short, justified text often results in awkward word spacing. This looks unprofessional and slows readers down. If you use a lot of hyphenation and good kerning, this may improve word spacing. However, hyphenated words are an additional factor that slows readers down even more (see also *Hyphenation—yes or no?* 73).

- Justified text requires more manual adjustments than left-aligned text. You're better off investing this time in improving your content.

Examples

✖ No:

This is a very narrow text column. If there are some long words, often a significant part of a line remains empty. Enabling hyphenation can make this problem less evident, but the readability of hyphenated words isn't good either.
Lorem ipsum dolor sit amet, consectetur adipiscing elit. Etiam risus eros, laoreet vulputate posuere vitae, luctus eget mi. Mauris cursus interdum urna eu interdum. Quisque et felis ipsum.

✔ **Yes:**

This is a very narrow text column. If there are some long words, often a significant part of a line remains empty. Enabling hyphenation can make this problem less evident, but the readability of hyphenated words isn't good either.

Lorem ipsum dolor sit amet, consectetur adipiscing elit. Etiam risus eros, laoreet vulputate posuere vitae, luctus eget mi. Mauris cursus interdum urna eu interdum. Quisque et felis ipsum.

✘ **No:**

This is a paragraph with very long lines of text (longer than they should be). Because all lines end at the same position, it's difficult to continue reading the correct next line. Try it for yourself: Lorem ipsum dolor sit amet, consectetur adipiscing elit. Etiam risus eros, laoreet vulputate posuere vitae, luctus eget mi. Mauris cursus interdum urna eu interdum. Quisque et felis ipsum. Praesent suscipit vulputate felis, non laoreet dui mattis at. Phasellus malesuada accumsan facilisis. Curabitur id convallis risus. Vestibulum eget mauris ullamcorper felis posuere laoreet. Integer eget quam at dui luctus blandit. Sed ac metus sed libero malesuada fermentum a nec odio. Integer vestibulum lobortis feugiat. Sed eu sapien arcu, a ultrices ante. Proin a neque velit, pharetra adipiscing dui. Quisque orci odio, dignissim et congue sed, condimentum et nulla.

✔ **Yes:**

This is a paragraph with very long lines of text (longer than they should be). Because all lines end at the same position, it's difficult to continue reading the correct next line. Try it for yourself: Lorem ipsum dolor sit amet, consectetur adipiscing elit. Etiam risus eros, laoreet vulputate posuere vitae, luctus eget mi. Mauris cursus interdum urna eu interdum. Quisque et felis ipsum. Praesent suscipit vulputate felis, non laoreet dui mattis at. Phasellus malesuada accumsan facilisis. Curabitur id convallis risus. Vestibulum eget mauris ullamcorper felis posuere laoreet. Integer eget quam at dui luctus blandit. Sed ac metus sed libero malesuada fermentum a nec odio. Integer vestibulum lobortis feugiat. Sed eu sapien arcu, a ultrices ante. Proin a neque velit, pharetra adipiscing dui. Quisque orci odio, dignissim et congue sed, condimentum et nulla.

2.2.5 Hyphenation—yes or no?

The advantages of hyphenation are:

- Hyphenation can sometimes make a paragraph one or two lines shorter.
- If you don't hyphenate, in narrow text columns and in narrow table columns, a significant portion of the line may remain empty if there are many long words in a sentence.

The disadvantages of hyphenation are:

- Hyphenation impairs reading fluency:
  - Hyphenated words can't be read as one unit.
  - When all lines of text end at approximately the same position, it's more difficult to move the eyes to the beginning of the next line without slipping into the wrong place.
- Well-placed hyphenation can mean a lot of extra work for the author.

General recommendations

- When in doubt, don't hyphenate.
- Never hyphenate in online help. Even if you use a fixed column width, different browsers might render the text differently. The same might happen on different operating systems or if different fonts are installed.
- Never hyphenate manually. Consider hyphenating only if your authoring tool provides high-quality automatic hyphenation for all languages in which your documents will be published.
- When hyphenating, configure your authoring tool so that it only hyphenates long words. To do so, tell the tool to avoid hyphenation if hyphenation leaves less than 5 characters at the end of the old line or at the beginning of the new line.
- Disable hyphenation completely for certain character styles (see *Recommended character styles* 219).

 (DE)

Ein Text in deutscher Sprache ist fast immer etwas länger als derselbe Text in englischer Sprache. In folgenden Fällen kann es daher sinnvoll sein, in der deutschen Version eines Texts Silben zu trennen, in der englischen Version jedoch nicht:

- wenn Sie Platz sparen müssen

- wenn in beiden Sprachen jeweils dieselben Inhalte auf derselben Seite stehen sollen; beide Dokumente haben dann exakt dieselben Seitenzahlen; der Support kann Benutzer auf eine bestimmte Seite verweisen, egal in welcher Sprache die Benutzer das Dokument lesen
- wenn Inhalte in einem mehrsprachigen Dokument parallel nebeneinanderstehen

Things that must never be hyphenated

Regardless of whether you decide to hyphenate in general, *never* hyphenate the following:

- company names
- product names
- parameter names
- function names
- lines of command syntax or source code
- headings

Turn off hyphenation for the corresponding character styles.

Related rules

Automate line breaks and page breaks 41

2.2.6   What page margins?

In printed manuals:

- Page margins should depend on how users will use the document and on what kind of visual impression you want to make.

- Use small page margins if you want to get as much information onto one page as possible. However, note that long lines of text can be very hard to read (see *What line length?* 95).

- Use large page margins if you want to make a more elegant visual impression or if it's likely that readers will add notes to the document.

- If you don't want to have particularly small or particularly large page margins, the left page margin plus the right page margin should take up about one third of the page. Also take into account the page size (see *What page size?* 61) and the body text font size (see *Which font size and font spacing?* 93).

In online help:

- When designing a template for online help, be aware of the fact that many users don't maximize help windows. For this reason, use a small page margin to save screen real estate.

- A good value for the page margin in online help is somewhere in between 5 pixels to 25 pixels.

General design principles

Make the top margin slightly smaller than the bottom margin (see also *Trust your visual judgment* 35).

If your layout provides a margin column, but you don't expect that there will be much text in this margin column, make the page margin next to the margin column smaller than usual.

If you use headers and footers, you must account for the total visual impression that they make. Trust your visual judgment. Mostly, you need to increase the size of a margin to provide room for a header or footer, but often you don't need to increase it for the full size of the header or footer.

Note:
The size of the page margins also depends on the chosen font, so you might have to fine-tune the margin in a second pass after choosing the font (see *Which font size and font spacing?* 93 and *What line length?* 95).

Print layouts with only right pages

Make the left margin, the top margin, and the right margin approximately identical, and make the bottom margin slightly larger. A good ratio is about 3:3:3:4 or 3:3:3:5. The absolute values depend on the paper size, the font size, and the visual impression that you want to make.

If you expect that readers will add notes, increase the size of the right margin.

If you expect that readers will print an electronically shipped manual to file it in a ring binder, increase the size of the left margin.

The following picture shows a typical right page.

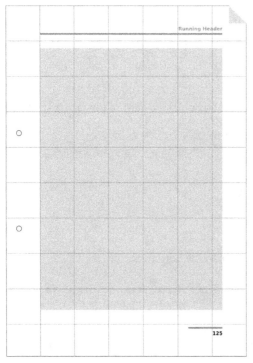

Print layouts with left pages and right pages

Make the inner page margins about half the size of the outer page margins. In total, you then get a visual impression of equal white space in the middle of the open manual and on either side. If the manual is going to have many pages, you need to slightly increase the inner page margins.

A good ratio typically is about 2.x : 3 : 4 : 5 (2.x is the inner margin, 3 is the top margin, 4 is the outer margin, and 5 is the bottom margin). The absolute values depend on the paper size, on the font size, and on the visual impression that you want to make.

The following picture shows a typical layout with left and right pages.

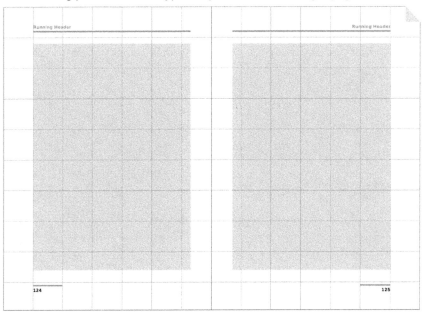

Another common way of finding a harmonious relation of margin widths is by using a true-to-scale sketch with diagonal lines. You can do so very comfortably with any vector-based image processor.

Depending on where you start on the first diagonal, you get type areas of different sizes.

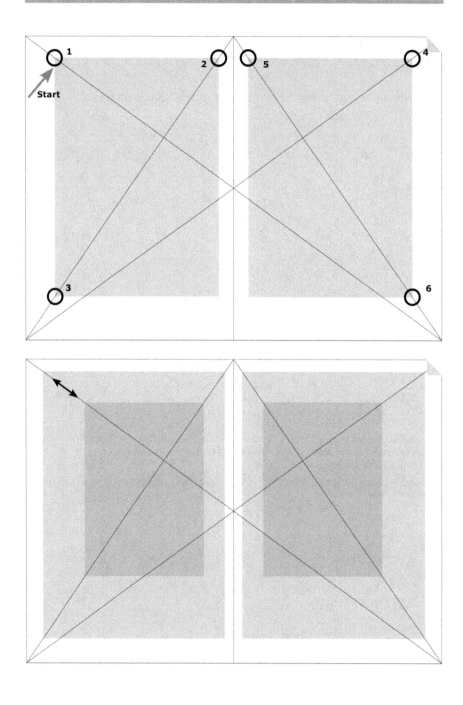

Related rules

▶ *What page size?* 61

Which font size and font spacing? 93

What line length? 95

2.2.7 Where to position headers, footers, and page numbers?

Headers and footers provide orientation:

- A running header shows the title of the current chapter or of the parent chapter, or it shows the titles of both the current chapter plus the parent chapter.

- Page numbers make it possible to access pages from the table of contents, from the index, or from cross-references.

- Don't add redundant information such as revision numbers, release dates, copyright notices, logos, and so on on each page (see *Use clear and simple design* 19).

When reading a page, readers focus on the text. They look at the header or footer only if they need orientation.

- Make the headers and footers visually unobtrusive.

- Nevertheless, make page numbers large enough so that they can easily be read when quickly thumbing through the book.

- Position page numbers so that they can easily be seen without having to open the book completely.

General guidelines

If you don't have any differing company-specific design guidelines that you need to follow:

- Put the running information about the current chapter title into the header rather than into the footer. This reflects the readers' decision chain: first check what the page is about, and then start reading or go to the next page.

- Choose the width and font size of the header so that even the longest of your headings won't be cropped.

- Clearly separate headers and footers from the body text. When possible, use enough white space to achieve the visual separation. If you have little space, use lines, but use a light color for the lines, preferably light gray.

- To reduce clutter, consider making a line shorter than the full page width. Often a short line is enough to indicate a separation.

- Reduce the visual weight of headers and footers by using a small font size or by using light colors, preferably a light shade of gray. You can combine a light shade of gray very well with a bold typeface.

  You can also use a narrower font—especially if space is too tight for long running headers. However, if you combine fonts, do so very carefully (see

Which font? 86).

- For documents that are to be printed double-sided, place the page numbers on the right side on right-hand pages and on the left side on left-hand pages. Don't center the page numbers in the middle of the page because there they can't be seen when you thumb through the book.

- If your pages have a small outer page border, align the page numbers to the outer edge of your body text.

- If your pages have a large outer page border, position the page numbers within the outer page border so that they can be seen easily even when the book isn't completely open.

- If you're going to make your document electronically available, always begin page numbering on the title page (you don't have to show the number). If you start with page 1 on another page, the page numbers on your pages will differ from the page numbers shown in the user's viewer, which can be very confusing.

If any regulations require you to include redundant information, such as logos, revision numbers, copyright information, and so on:

- Make the redundant elements as small as possible.

- Use light colors.

- In a printed document, place the redundant elements near the bottom of the page, where they're less prominent.

- In online content, place the redundant information in the lower right corner (see also *Topic area layout* 124).

Examples

The following pictures show some typical examples of headers and footers for your inspiration:

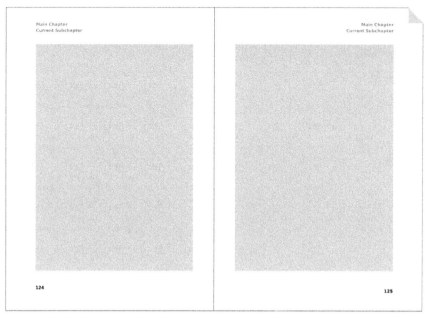

Main Chapter
Current Subchapter

Main Chapter
Current Subchapter

124

125

Running Header

Running Header

124

125

2.3 Choosing fonts and spacing

The chosen fonts and the font settings play a key role in the image that your document conveys.

Fonts and front sizes also have a major effect on readability.

Tip:
If you're unsure about a specific setting for a printed manual, don't hesitate to print a test page on your office printer. On paper, the visual appearance and readability are often very different than they are on screen. Put several alternatives next to each other and compare them. When in doubt, ask a second person. This person doesn't necessarily have to belong to your audience. Questions of readability are quite universal.

Key questions

Choosing fonts and setting font attributes involves the following questions:

- *Which font?* 86
- *Which font style?* 91
- *Which font size and font spacing?* 93
- *What line length?* 95
- *What line spacing?* 98

Related rules

Use clear and simple design 19
Recommended character styles 219

2.3.1 Which font?

If your company has design guidelines that require you to use a particular font, do so. If you ship your documents electronically, make sure that all users have the required fonts. If you have the proper license, embed the fonts, or ship them along with your product.

If your company doesn't have any corporate fonts, use a common font that's easily readable. Don't choose any stylish, trendy fonts. If you ship your documents electronically, use a font that's installed on the computers of all users. Bear in mind that the default fonts aren't identical on all operating systems.

Use as few fonts as possible (see *Use clear and simple design* 19).

Types of fonts

There are three main types of fonts:

- serif fonts
- sans-serif fonts
- monospaced fonts

This is a **serif font**.
The small embellishments at the
end of the letters are called "serifs."

This is a **sans-serif font**.
"Sans" is French and means "without."
So, this is a font without serifs.

This is a **monospaced font**. Fonts like this were common on mechanical typewriters.
All letters have the same width. For example, the "i" is just as wide as the "m." As a result, all characters are precisely positioned below each other. The space character also has the same width as all other characters.

(DE)

Deutsche Begriffe: serifenlose Schrift, Serifenschrift, nichtproportionale Schrift

General recommendations

In general:

- Sans-serif fonts tend to look quite modern and innovative, whereas serif fonts look more traditional. Using one or the other can emphasize the image of your product into an intended direction.
- Sans-serif fonts usually have better readability—particularly on screen. For this reason, always use sans-serif fonts in online help.
- The serifs of serif fonts guide the eye along the line. This makes reading long texts slightly less tiring and is the reason why most novels have serif fonts. In user assistance, however, users usually read only short passages, so this advantage is rather unimportant here.
- Don't use monospaced fonts except for source code and user input.
- Avoid all sorts of squiggly fonts.
- Avoid fonts that have very thin lines. Many people have trouble reading them. On screen, thin lines may disappear completely.
- Avoid fonts that have very thick lines. They make your whole text look bold and obtrusive, and you lose the possibility of making individual words bold. In addition, fonts with thick lines need a lot of space, so you can put less information onto one page.
- Avoid fonts that have lowercase letters with a very small height. This is particularly important in online help, where the height of lowercase letters should be at least 50% of the height of capital letters to be clearly readable.
- If you ship your document electronically, make sure that the font is available on the users' computers, or make sure that you can embed the font into the document.
- Make sure that the fonts that you choose support all language-specific characters of the languages into which your document might be translated. Plan ahead. As business evolves, languages that may seem unlikely to be used today may soon become mission-critical.

Tip:
When in doubt, use the more common font. Most users find reading a familiar font more relaxing than reading an unfamiliar one.

Large and small fonts

In places where space is tight, you sometimes need to add very small text. However, when scaled down, the readability of many standard fonts is very poor. If your standard font looks blurred in cases like this, try a font that has taller lowercase letters. When possible, look for a font that's been particularly designed to be used in small sizes.

Likewise, many fonts aren't appropriate for large headings either. If a font looks inept here, try a narrower font or a font that has a smaller line width.

Combining fonts

In general, use as few fonts as possible.

Use an additional font only if this improves the readability of your document. This typically is the case if the additional font marks a special element, such as headings, labels, or computer source code. Also, you sometimes need a special font for extremely small text or for extremely large text.

The key rules for combining fonts are:

- Avoid mixing more than two fonts. Note that if you have a company logo or product logo on a page, there may be yet another font in the logo.

- Don't mix fonts that look very similar. Some readers won't notice; others might think that it's a mistake.

  However, do use fonts that have *some* characteristics in common so that the combination looks intentional. For example, combine two fonts that have equal line widths, two fonts that are both modern, two fonts that are both elegant, and so on.

- Use fonts that have identically sized lowercase letters.

- Finding a good combination of two serif fonts is especially difficult. Mixing two sans-serif fonts tends to be easier.

- A good option can be to mix a serif font with a sans-serif font. Often, for example, a serif font is used for body text, and a sans-serif font is used for headings. For some fonts, there's even both a serif version and a sans-serif version available, which blend especially well.

Bold and italic font styles

The standard function in many editors that makes text bold or italic does so by applying a mathematical algorithm to the used typeface. This usually produces acceptable results.

However, if you want to achieve a truly professional layout in a printed manual, don't apply the automatic functions; instead, assign a special bold version or italic version of the font that you use. For most professional fonts, special bold and italic versions are available, but they must often be purchased separately.

> ❶ **Important:** For online help, always use the standard bold and italic function. If you use a special font version, this version might not be available on the users' computers.

Recommended fonts

Avoid using Arial in body text because this font is very narrow and thus often results in poor readability and in an amateurish look. However, the fact that Arial is very narrow makes it an excellent choice for headings.

Some good, easily available fonts for the body text of printed manuals are:

- **Bitstream Vera** and **Bitstream Vera Sans** (initially developed for Linux; quite similar to Verdana)
- **Calibri** (sans-serif; ships with Windows)
- **Corbel** (sans-serif; ships with Windows)
- **Consolas** (monospaced; a good alternative to Courier; ships with Windows)
- **Constantia** (serif; designed to be used both on paper and on screen; ships with Windows)
- **Meta** (must be licensed separately)
- **Univers** (must be licensed separately)
- **Syntax** (must be licensed separately)

Common fonts for online help that are installed on most computers are:
- **Arial** (ships with Windows)
- **Verdana** (ships with Windows)
- **Segoe UI** (ships with Windows)
- **Trebuchet MS** (ships with Windows)
- **Courier New** (monospaced; ships with Windows)

In online help, generally use Verdana. Verdana has an excellent readability on screen and is installed on most computers. Use Arial only if you have little space—for example, in headings and in narrow table cells. Use Courier New for program source code.

In HTML, use the font-family attribute to specify some replacement fonts. If a particular font isn't available on a user's computer, the browser then uses the replacement font instead of the original font. (If you don't specify a replacement font, the browser uses its default font, which may be very different from your original font.)

If your authoring tool doesn't support adding the font-family attribute, consider running a global search and replace on the final HTML documents.

For Verdana, the font-family attribute may look like this:
```
font-family: 'Verdana', 'Bitstream Vera Sans', 'Trebuchet MS',
'Geneva', 'Arial', sans-serif;
```

Examples

Bitstream Vera	How do you like this?
Calibri	How do you like this?
Corbel	How do you like this?
Consolas	How do you like this?
Constantia	How do you like this?
Meta	How do you like this?
Univers	How do you like this?
Syntax	How do you like this?
Arial	How do you like this?
Verdana	How do you like this?
Segoe UI	How do you like this?
Trebuchet MS	How do you like this?
Courier New	How do you like this?

Related rules

Which font style? 91
Which font size and font spacing? 93

2.3.2 Which font style?

Avoid variety (see *Use clear and simple design* 19).

The fewer font styles you need, the better.

Use only one font attribute at a time. For example, make text bold, *or* italic, *or* underlined, but don't make it bold + italic + underlined.

Font style Bold

Bold text should *identify* information.

Make headings bold.

Within the body text, only make bold what you want to use as a visual label to support skimming.

Font styles Italic and Small Caps

It's harder to read italic text than to read normal text, especially when reading on screen.

Avoid long text in italics; use italics only to *emphasize* individual words or phrases.

Don't use small capitals at all. Their readability is very poor *both* in printed manuals *and* in online help.

 (DE)

Deutscher Begriff: Kapitälchen

Font style Underlined

In online help, use underlined text exclusively for hyperlinks. Even if links are not underlined in your document, don't use underlined text for anything else; many users will click the underlined text because they think that it's a link.

If you've underlined text in a printed manual, only use it for single words or short phrases. Long sections of underlined text are very difficult to read. What's more, if you underline a long section, the underlining loses its function of pointing out what's *especially* important.

Examples

This is some regular text.

This is some **bold** text.

This text is *italic*.

This text is <u>underlined</u>.

Don't use SMALL CAPS.

Don't use ALL CAPS.

<u>Don't combine</u> font attributes.

In online help, use underlined text for <u>links</u> only.

It's hard to read a long italic text. Try it for yourself: Lorem ipsum dolor sit amet, consectetur adipiscing elit. Aenean ac orci quis massa vestibulum mollis. Sed rutrum purus ac felis blandit sed sodales justo auctor. Suspendisse mi lectus, lobortis vitae consectetur vel, facilisis id nisi. Duis nulla ligula, interdum vulputate suscipit id, fringilla sed nisi. Nam vel lacinia libero. Vivamus sagittis sapien in nibh dapibus a fermentum ligula placerat. Pellentesque dolor enim, volutpat quis ultricies quis, vehicula nec nisi. Sed ornare pellentesque odio eget mattis. Ut fermentum pulvinar nisi, at sodales felis fermentum ac.

Related rules

Which font? 86

Which font size and font spacing? 93

Recommended character styles 219

2.3.3 Which font size and font spacing?

Don't use a poorly readable font size for the sake of design. In user assistance, readability is more important than beauty.

Keep in mind:

- Many people have poor eyesight. If *you* can read something well, this doesn't mean that *others* can also read it well.
- Your document may be read in places or under conditions where reading is more difficult than in your office.
- Audiences that aren't used to reading long texts may be discouraged by small fonts.
- Type smaller than 10 points usually slows down *all* readers.

General recommendations

Adequate font sizes for printed manuals usually are:

- on small paper sizes: body text from 8.5 points up to 10 points, headings up to 20 points
- on large paper sizes such as A4 or Letter: body text from 10 points up to 12 points, headings up to 24 points
- if the manual is likely to be read under difficult conditions: body text up to 14 points

Adequate font sizes for online help usually are:

- 10 points or 11 points for body text
- 14 points for headings

Other factors that influence font size:

- In texts that are mainly used for looking up small pieces of information, eye exhaustion can be neglected. So you can use a smaller font size here. Dictionaries are a typical example of this.
- Also base your decision on the specific font that you're using. Not all fonts of a given font size actually have the same character size. For example, in one font with a font size of 10 points, letters may be 2.5 mm high, whereas in another font with a font size of 10 points, letters may only be 2.0 mm high.

Font size in headings

The font size is the most important way to visualize the hierarchy of headings (other ways are the font weight, color, and space before and after the heading; see *Heading paragraph styles* 187).

- The size runs from biggest to smallest, and the typeface runs from boldest to lightest.

- In a printed manual, use at least a difference of 3 points between the levels of headings (for example, 20, 17, 14 points). Otherwise, readers might be unsure of what level they're looking at.

- In online help, topics are more independent. Here, use the same font size for all topic titles, regardless of where they're located within the hierarchy of the table of contents.

Optimizing font spacing for large font sizes

For font sizes up to 16 points, the default font spacing usually provides the best results.

If you use font sizes that are larger than 16 points (mainly in headings), you often need to decrease the default font spacing for better readability. As a positive side effect, this also makes the headings shorter and reduces the risk of having unwanted line breaks.

If you use a light font color on a dark background (inverted text), increasing font spacing usually also improves readability.

In online help, slightly increased font spacing can improve readability as well, especially if you don't use a font that's optimized for reading on screen.

Related rules

Which font? 86

What page size? 61

What page margins? 75

What line length? 95

2.3.4   What line length?

The number of characters that can be printed within one line depends on:

- the page size (see *What page size?* 61)
- page margins (see *What page margins?* 75)
- the font (see *Which font?* 86)
- the font size (see *Which font size and font spacing?* 93)

All of these settings should work together to achieve an adequate line length.

If lines are too long:

- When reading, moving the eyes isn't enough. Readers must also move their heads.
- You don't have an overview of the whole line. This makes it difficult to go to the beginning of the next line without slipping into the wrong line.

If lines are too short:

- Sentences are frequently interrupted by line breaks, which slows down reading.
- The layout looks amateurish:
  - If you hyphenate, there are many hyphenated words.
  - If you don't hyphenate, there's a lot of unused white space.

Recommendations

Adequate line lengths are roughly in between:

- 40 to 70 characters per line
- 8 to 12 words per line (this value applies to English; most other languages have slightly longer words, so a number of 7 to 11 words is more adequate there)

Use line lengths near the upper limit of 12 words per line if:

- your readers are well-educated
- your text has many long paragraphs
- you use a serif font

Examples

In the following example, the text has an average of approximately 30 characters. This is clearly too short:

Now test the readability of different line lengths. You will notice that reading is harder when lines are too short or too long. Now let's make things a little harder. Here comes some Latin text: Lorem ipsum dolor sit amet, consectetur adipiscing elit. Ut faucibus dignissim mattis. Nullam ut lobortis augue. Nulla viverra, elit semper gravida tempor, nulla risus luctus tellus, quis iaculis metus sem sed purus.

An average of approximately 60 characters is best:

Now test the readability of different line lengths. You will notice that reading is harder when lines are too short or too long. Now let's make things a little harder. Here comes some Latin text: Lorem ipsum dolor sit amet, consectetur adipiscing elit. Ut faucibus dignissim mattis. Nullam ut lobortis augue. Nulla viverra, elit semper gravida tempor, nulla risus luctus tellus, quis iaculis metus sem sed purus. Nullam porttitor sagittis interdum. Cras porta lobortis neque, a pretium libero suscipit sed. Pellentesque habitant morbi tristique senectus et netus et malesuada fames ac turpis egestas. Donec libero sapien, gravida id interdum ut, egestas at erat. Sed non turpis erat. Morbi tristique scelerisque ultricies.

An average of approximately 75 characters is already too long:

Now test the readability of different line lengths. You will notice that reading is harder when lines are too short or too long. Now let's make things a little harder. Here comes some Latin text: Lorem ipsum dolor sit amet, consectetur adipiscing elit. Ut faucibus dignissim mattis. Nullam ut lobortis augue. Nulla viverra, elit semper gravida tempor, nulla risus luctus tellus, quis iaculis metus sem sed purus. Nullam porttitor sagittis interdum. Cras porta lobortis neque, a pretium libero suscipit sed. Pellentesque habitant morbi tristique senectus et netus et malesuada fames ac turpis egestas. Donec libero sapien, gravida id interdum ut, egestas at erat. Sed non turpis erat. Morbi tristique scelerisque ultricies.

Related rules

What page size? 61

What page margins? 75

Which font size and font spacing? 93

2.3.5   What line spacing?

Line spacing is the vertical distance between two lines in the same paragraph.

First line of the paragraph.

Second line of the *same* paragraph.   **line spacing**

The disadvantage of large line spacing is that the text requires more space, which means that you can include less information on a page. In addition, if line spacing is too large, the paragraph loses its cohesion.

The advantage of large line spacing is that it helps readers to stay in the right line while reading, especially while they move the eyes from the end of a line to the next line.

Recommendations

General formula:

- For printed manuals, adequate line spacing is roughly about 120% of the font size. For example, if you use a font size of 10 points, use a line spacing of approximately 12 points.

- In online help, line spacing should be slightly higher than in printed manuals. A good rule of thumb is about 120% to 150% of the font size, sometimes even more.

Additional factors:

- The larger the font is, the smaller the relative line spacing should be. This especially applies to headings. For example, if you use a font size of 48 points, a line spacing of 44 points is enough.

- Long lines need larger line spacing than short lines because with long lines the need to guide the readers' eyes is more important than with short lines.

- Sans-serif fonts tend to need slightly more line spacing than serif fonts need because with serif fonts the serifs already do part of the job of guiding the readers' eyes.

Examples

The following example uses a line spacing of 100%. That's too small for body text.

Test the readability of different line spacings. You'll notice that reading is harder when there is less space. When there is more space, paragraphs lose their cohesion. Here is some Latin text, so you can now fully concentrate on the visual appearance rather than on the content: Lorem ipsum dolor sit amet, consectetur adipiscing elit. Ut faucibus dignissim mattis. Nullam ut lobortis augue. Nulla viverra, elit semper gravida tempor, nulla risus luctus tellus, quis iaculis metus sem sed purus. Nullam porttitor sagittis interdum. Cras porta lobortis neque, a pretium libero suscipit sed. Pellentesque habitant morbi tristique senectus et netus et malesuada fames ac turpis egestas. Donec libero sapien, gravida id interdum ut, egestas at erat. Sed non turpis erat. Morbi tristique scelerisque ultricies.

The next example uses a line spacing of 120%. This is generally OK for body text.

Test the readability of different line spacings. You'll notice that reading is harder when there is less space. When there is more space, paragraphs lose their cohesion. Here is some Latin text, so you can now fully concentrate on the visual appearance rather than on the content: Lorem ipsum dolor sit amet, consectetur adipiscing elit. Ut faucibus dignissim mattis. Nullam ut lobortis augue. Nulla viverra, elit semper gravida tempor, nulla risus luctus tellus, quis iaculis metus sem sed purus. Nullam porttitor sagittis interdum. Cras porta lobortis neque, a pretium libero suscipit sed. Pellentesque habitant morbi tristique senectus et netus et malesuada fames ac turpis egestas. Donec libero sapien, gravida id interdum ut, egestas at erat. Sed non turpis erat. Morbi tristique scelerisque ultricies.

Now, here's an example that uses a line spacing of 150%. This is a good choice if lines are long, especially in online help.

Test the readability of different line spacings. You'll notice that reading is harder when there is less space. When there is more space, paragraphs lose their cohesion. Here is some Latin text, so you can now fully concentrate on the visual appearance rather than on the content: Lorem ipsum dolor sit amet, consectetur adipiscing elit. Ut faucibus dignissim mattis. Nullam ut lobortis augue. Nulla viverra, elit semper gravida tempor, nulla risus luctus tellus, quis iaculis metus sem sed purus. Nullam porttitor sagittis interdum. Cras porta lobortis neque, a pretium libero suscipit sed. Pellentesque habitant morbi tristique senectus et netus et malesuada fames ac turpis egestas. Donec libero sapien, gravida id interdum ut, egestas at erat. Sed non turpis erat. Morbi tristique scelerisque ultricies.

The next example uses a line spacing of 200%. That's too large.

Test the readability of different line spacings. You'll notice that

reading is harder when there is less space. When there is more

space, paragraphs lose their cohesion. Here is some Latin text, so

you can now fully concentrate on the visual appearance rather

than on the content: Lorem ipsum dolor sit amet, consectetur

adipiscing elit. Ut faucibus dignissim mattis. Nullam ut lobortis

augue. Nulla viverra, elit semper gravida tempor, nulla risus

luctus tellus, quis iaculis metus sem sed purus. Nullam porttitor

sagittis interdum. Cras porta lobortis neque, a pretium libero

suscipit sed. Pellentesque habitant morbi tristique senectus et

netus et malesuada fames ac turpis egestas. Donec libero sapien,

gravida id interdum ut, egestas at erat. Sed non turpis erat.

Morbi tristique scelerisque ultricies.

With large fonts, the relative line spacing must be smaller than with small fonts. Other than with the smaller font in the first example, here a line spacing of 100% is enough:

Testing the Readability in Multiline Headings

In a large heading like this, a line spacing of 120% is too much. The heading loses its cohesion and looks like two separate headings instead of one heading:

Testing the Readability in Multiline Headings

2.4 Recommended screen layouts

The screen layout must allocate room for:

- a table of contents (see *Table of contents area layout* 109)
- an alphabetical index (see *Index area layout* 118)
- full-text search (see *Search area layout* 121)
- the topic contents (see *Topic area layout* 124)

Two-pane layout

Traditionally, most online help systems consist of two panes: a navigation area and a separate topic area.

The navigation area is typically positioned on the left side, and the topic area on the right side. For help systems written in left-to-right languages, this order matches the logical sequence of operations: first choose, and then read.

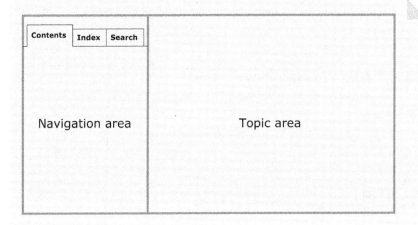

Methods of switching the view within the navigation area

Within the navigation area, users can typically switch the view between a table of contents, an alphabetical index, and full-text search.

The following picture contrasts the most common implementations:

- horizontal tabs
- vertical tabs
- expandable sections

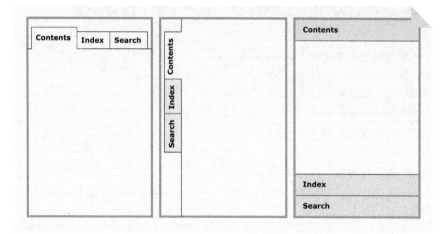

 (DE)

Standardbezeichnungen in deutschsprachigen Hilfen: *Inhalt, Index, Suche*

Methods of visually separating the areas

To separate the navigation area clearly from the topic area, you can use:

- a fine line
- a light background color on the navigation area

Note:
Don't use the background color on the topic area. Because the topic area is larger than the navigation area, a colored background would be too dominant here. In addition, a dark background might impair readability, which is more serious in the topic area.

A good solution is to use a gradient instead of a solid background color. A gradient is less obtrusive but provides enough contrast on the borderline between both areas.

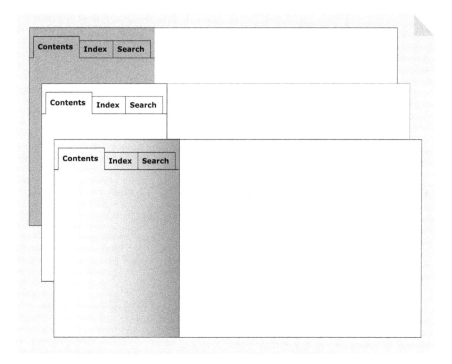

How wide should the navigation area be?

Some help authoring tools and help viewers allow you to set the default width of the navigation area.

- Make the navigation area as wide as necessary but as small as possible. In particular, don't make it wider than needed for the longest entry in the table of contents.

- If your help will be translated: Bear in mind that in most other languages, texts are slightly longer than in English.

- On a standard computer monitor, the relative width of the navigation area should typically be in between 25% and 33% of the help window's total width.

Navigation areas that can be hidden

Some help systems hide the navigation area by default, especially if there's very little screen real estate available or if help is embedded into the program window as an integrated help pane. Users can then use a special button to toggle the display of the navigation area.

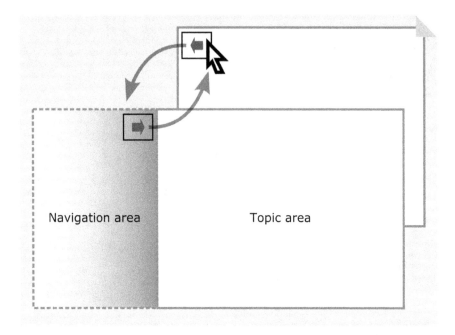

The advantages of a hideable navigation area are:

- When the navigation area is hidden, the help window needs less screen real estate.
- When the navigation area is hidden, users can fully focus on the content of the topic. They are less tempted to go to other topics that they see in the table of contents.

The disadvantages of a hideable navigation area are:

- Hiding and unhiding requires additional interaction.
- When the navigation area is hidden, users don't see the global context into which the topic that they're reading is embedded.
- When the navigation area is hidden, users don't find related information unless there are links to this information in the topic.
- Novice users might be completely unaware that they can open an extra navigation pane—even if the button can easily be seen.

Layouts that don't have a separate navigation area

Help on the Internet and on mobile devices often doesn't use a separate navigation area at all but switches between the topic view, the table of contents, the index, and search results. Only the search box is typically available on all pages.

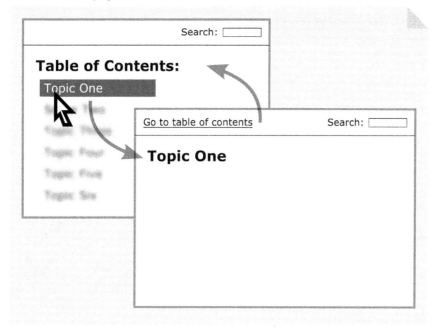

The advantages of switching between content and navigation are:

- The help window only needs a minimum of screen real estate.
- When navigation isn't visible, users aren't tempted to go to other topics that they see in the table of contents. Instead, users can fully focus on the content of the topic.
- Compared to a hideable navigation area, this approach is more consistent and less confusing for novice users. The size of the help window and the size of its panes always remain constant.

The disadvantages of switching between content and navigation are:

- Switching requires additional interaction.
- When navigation is hidden, users don't see the global context into which the topic that they're reading is embedded.
- When navigation is hidden, users don't find related information unless there are links to this information in the topic.

Which basic screen layout should you use?

If you have a lot of content, and if you don't have to embed your content into an existing web site, use the traditional version that places the navigation area and the topic area side by side.

Make the table of contents hideable or hide it by default only if you have very little screen real estate, particularly if the help window is embedded into a program window.

Use the model of switching between different views for the topics, the table of contents, the index, and search results if:

- you only have a small number of help topics
- you need to embed online help into an existing web site
- help will be used on devices that have a small screen size, such as mobile devices

Note:
Some help authoring tools don't support all types of layouts.

Related rules

Setting the type area 59

Recommended page layouts 128

2.4.1 Table of contents area layout

The electronic table of contents typically consists of a hierarchical list of topics. Topics that have subtopics are commonly represented by book symbols. Topics that don't have subtopics are represented by page symbols.

Most help authoring tools create a well-designed hierarchical table of contents automatically. Yet, often you can significantly improve the usability of online help.

- You can use custom icons that match the information types of topics.
- You can optimize how users can interact with the table of contents.

In addition to the classical hierarchical table of contents, also consider using or adding alternative forms of tables of contents.

General design tips and examples

Fonts:

- If space is limited, use a narrow font for the table of contents. If you use Verdana for the topic text, for example, Arial is a good narrower choice for the table of contents. Use the same font size as the body text within the topic area.
- Unlike in the table of contents of a book, use the same font and the same font style for all hierarchy levels.

Indentation and symbols:

- Indent subtopics.
- If you can configure it, don't use plus-minus icons and connecting lines in front of the topic icons. They add little value and lots of clutter.

Color:

- Don't use bold colors for the topic icons because they don't contain important information. Instead, use fairly pale colors or shades of gray.
- Likewise, don't use a bold color to highlight the selected topic. Inverted text (white on dark gray) is perfectly good enough for highlighting the selected topic. Reserve color for highlighting important information in your help content. If you don't like a black-and-white table of contents, use one basic color that you already use somewhere else. For example, use your corporate color, the predominant color of your product logo, or the color used in the tab that brings up the table of contents.

✖ No:

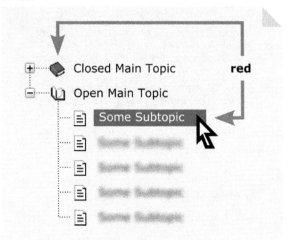

✔ Yes:

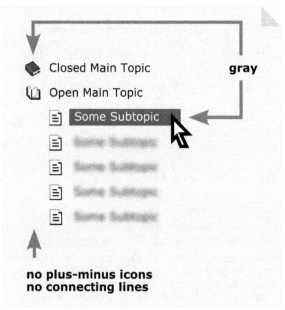

**no plus-minus icons
no connecting lines**

Using different topic icons (optional)

If your topics match particular information types, you can use a dedicated topic icon for each information type. Readers can then see immediately from the topic icon which kind of information a topic contains before they even read the heading.

In the topic icons, you can use colors, symbols, or a combination of both.

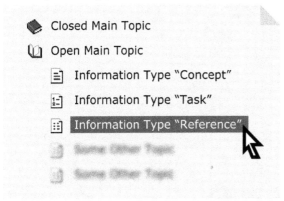

Instead of indicating the information type, you can also use special topic icons to indicate:

- new topics
- topics of particular importance
- topics for particular user groups

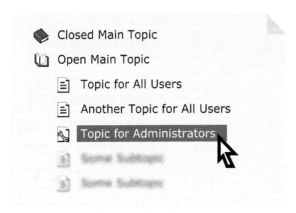

Even a small difference in the icon design can give a hint:

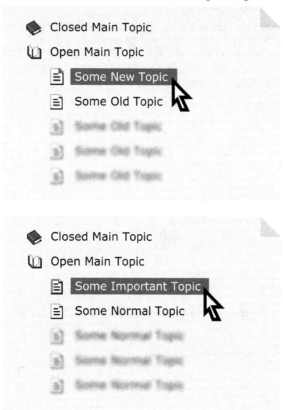

Should topics open with a single click or with a double click?

When possible, configure your authoring tool so that topics can be opened with a single click rather than with a double click. A single click better matches what users are used to from surfing the Internet.

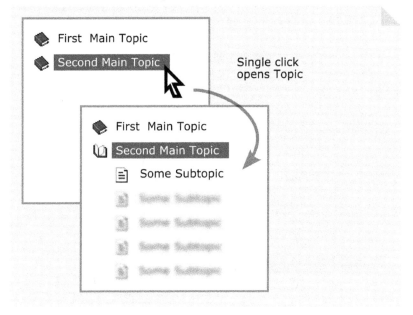

Should an entry collapse when another entry is opened?

Some help authoring tools allow you to configure whether an entry collapses when a user jumps to another topic. The advantage of having the previous entry collapse is that users always only see a small part of the whole structure, which helps them to focus on the section that they're currently browsing. The downside is that they can't see the complete structure all at once.

As rules of thumb:

- In very long documents, make entries automatically collapse.
- In short documents and in medium-sized documents, allow multiple chapters to be expanded simultaneously.

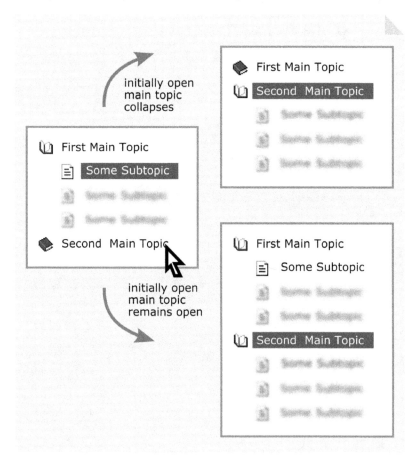

Should the table of contents be synchronized with the topic view?

When possible, enable synchronization of the table of contents with the topic that's currently shown. When a user jumps to another topic via a link, via search, or via the index, the table of contents then automatically expands at that position and highlights the newly selected topic.

- The advantage of synchronization is that users can constantly see their current position within the table of contents, which improves orientation.

- The disadvantage of synchronization is that users need to use the Back button if they want to return to the topic from where they've started. In the early days of the Internet, this was a problem because many users didn't know how to use the Back button. Nowadays, however, this usually isn't a problem, so in most cases enabling synchronization is the better alternative.

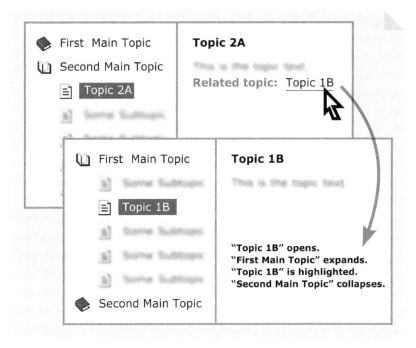

Alternative or additional forms of a table of contents

A table of contents doesn't necessarily have to look like the ordinary collapsible table of contents.

If the table of contents is short, its items don't have to be collapsible. This works especially well on mobile devices with small displays. In its most simplistic form, a table of contents can just be a static list of topics.

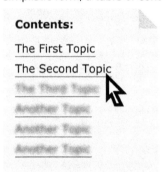

Depending on the subject of your document, you can sometimes use a good visual metaphor relating to physical objects or workflows. Occasionally, you can design a table of contents as a flowchart, or as an organizational chart.

The visual table of contents can replace the classical table of contents completely, or you can use it in addition to the classical table of contents. This is one of the few opportunities in technical communication where you can really be creative.

For your inspiration, the following pictures show some examples of "tables" of contents that use a visual metaphor:

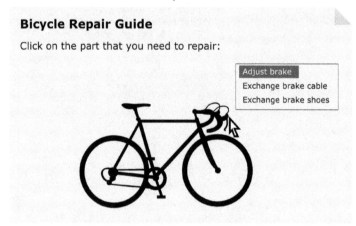

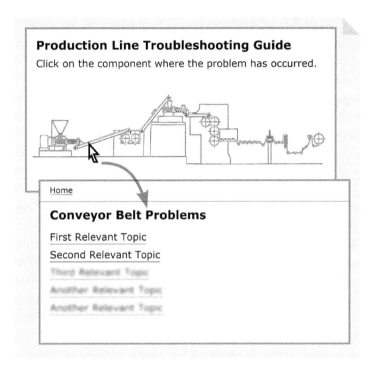

Production Line Troubleshooting Guide

Click on the component where the problem has occurred.

Home

Conveyor Belt Problems

First Relevant Topic

Second Relevant Topic

 Related rules

Recommended screen layouts 103

2.4.2 Index area layout

Most help authoring tools create a well-designed index automatically.

- If you create a compiled help format, usually there aren't any customization options.
- If you create browser-based help (plain HTML), the only thing that you usually need to do is to customize fonts and colors.

The key elements of the index area are:

- a search box where users can enter the first letters of the term to search for; alternatively, a list of letters with links that take users to the places where the index entries for each letter start
- subheadings for each letter
- index entries
- index subentries

Design tips and examples

When possible, put the search box or the links into a non-scrolling region so that they remain in place when users scroll down within the index.

If you add letters and links, omit those letters in the index for which there's no entry.

If space is limited, use a narrow font. If you use Verdana for the topic text, Arial is a good choice for the index entries. Use the same font size as for the topic text.

Within the list of index entries, make the subheadings for each letter visually prominent so that readers can identify each section clearly even while scrolling. Use bold font style and an increased space before the paragraph. Indent subentries.

 (DE)

Behandeln Sie Umlaute nicht gesondert. Behandeln Sie *Ä* wie *A*, *Ö* wie *O* und *Ü* wie *U*.

- Fügen Sie also *keine* zusätzlichen Abschnitte für die Umlaute ein.
- Sortieren Sie ein *ä* so ein wie ein *a* (nicht wie *ae*), ein *ö* so wie ein *o* (nicht wie *oe)* und ein *ü* so wie ein *u* (nicht wie *ue*).

Behandeln Sie ein *ß* wie *ss*.

The following picture contrasts some different implementations:

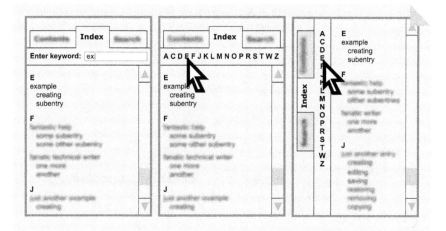

Synchronization with the table of contents

If your authoring tool supports it, activate synchronization with the table of contents so that when a user clicks an index entry the corresponding topic is also highlighted in the table of contents.

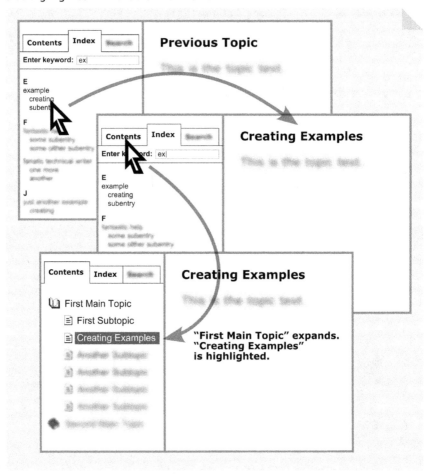

Related rules

Recommended screen layouts 103

2.4.3   Search area layout

Most help authoring tools create a well-designed search area automatically.

- If you create a compiled help format, usually there aren't any customization options.
- If you create browser-based help (plain HTML), the only thing that you usually need to do is to customize fonts and colors.

The key elements of the search area are:

- a search box where users can enter the term to search for
- a search button that triggers the search operation
- a list of search results linking to the found topics

Design tips and examples

Near the search box, add some hints that explain how to use the search function and which advanced search options are available.

If space is limited, use a narrow font. If you use Verdana for the topic text, Arial is a good choice for the search results. Use the same font size as for the body text within the topic area.

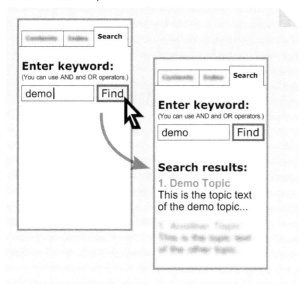

Highlighting search terms

If your authoring tool and the used search engine support it, you can activate the highlighting of search terms within the topic text. This feature helps users to find the relevant spots more quickly. There is no significant disadvantage, so *always* enable this feature.

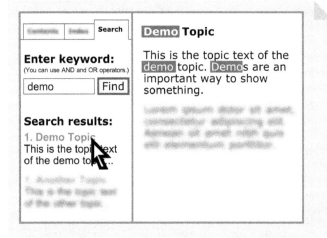

Synchronization with the table of contents

If your authoring tool supports it, activate synchronization with the table of contents so that when a user clicks a found topic, that topic is also highlighted in the table of contents.

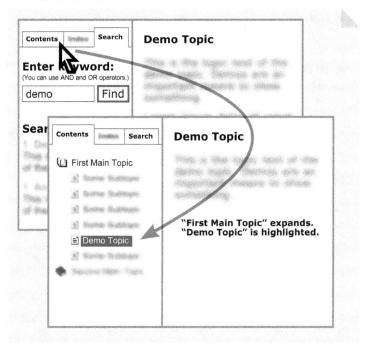

"First Main Topic" expands.
"Demo Topic" is highlighted.

 Related rules

Recommended screen layouts 109

2.4.4 Topic area layout

The key elements of the topic area are:

- the topic title
- optional: a breadcrumb trail, the links of the browse sequence, and a print button plus other buttons
- the topic text
- optional: related topic links
- optional: a footer with version information, copyright information, and a logo (not recommended)

Design tips and examples

Background:

- For optimum readability, choose white or a very light color for the background color.
- Never add background textures or background pictures. This gives your help a truly amateurish look and seriously impedes readability. Text displayed on textures or pictures is very hard to read.

Type area:

- Many users don't maximize help windows. For this reason, use a small page margin to save screen real estate. A good value for the page margin is somewhere in between 5 and 25 pixels.
- To minimize the users' need to scroll, in online help always use only one column. If you want to align pictures side by side with instructions, you can do so locally within a topic with the help of a borderless table.

Topic title:

- In online help, use the same heading style for *all* topics, no matter where they stand in the hierarchy of the table of contents.
- To clearly separate the heading from the topic text, add sufficient space below the heading. Alternatively, add a thin line, which can save some screen real estate but adds more clutter. If you add a line, try to align it with one or more elements from the navigation area (see *Align texts and objects to a design grid* 30).
- For details on choosing fonts, see *Choosing fonts and spacing* 85 . For details on setting up the required styles, see *Recommended paragraph styles* 179 and *Recommended character styles* 219 .

Additional navigation tools:

- In browser-based help, a good place for buttons, such as a print button or browse buttons, is at the right side of the heading. Align the heading to the left and the buttons to the right.

- In compiled help, most buttons are already integrated into the help viewer's toolbar, so you don't have to add anything here. Don't add a redundant copy of a button that already exists.

- If you add a breadcrumb trail, the best place to do so is above the heading. This matches the sequence that readers must follow to access the topic.

- If you provide related-topic links, add them at the end of the topics. Various studies have shown that most users prefer having related-topic links at the end of the topic rather than at the beginning—even if this means that users must scroll to view the links.

- If you can, avoid having a footer, which mostly just adds clutter to the page without providing any real value to the user.

The following pictures show some typical topic area layouts with and without a breadcrumb trail and with and without lines:

Handling logos and copyright notices

If you can, avoid adding redundant information to each topic. Typical examples of redundant information are the product name, the document name, the version number, the copyright notice, and logos.

The proper place for the product name, the document name, and the version number is the title bar of the help window.

If your corporate regulations oblige you to add a logo or other items to each topic, try to take as much visual weight away from these elements as possible:

- Add a fine bottom line below the topic that separates the "real" content from the redundant elements.

- Use a small font and a light color, preferably a light shade of gray.

- Place the most eye-catching element (usually the logo) into the lower right corner.

Should you make the heading non-scrolling?

Some authoring tools and help viewers let you decide whether the heading moves along with the topic text when users scroll up and down or whether it's "non-scrolling" and stays in place at the top of the topic area in the help window.

- The advantage of a non-scrolling heading is that users can constantly see what the topic that they are currently reading is about. If you have the breadcrumb trail, browse buttons, or other buttons in this area, they also remain in place.

- The disadvantage of a non-scrolling heading is that it always takes up some screen real estate. As a result, users need to scroll more.

From surfing the Internet, most users are used to headings that scroll away, so standard headings that scroll away are usually the better solution also in online help.

Non-scrolling headings can be a good choice when the heading itself contains some important information. A typical example of this is a command reference where the heading shows the syntax of the command and its parameters.

Related rules

Recommended screen layouts 103

2.5 Recommended page layouts

Create a consistent look in the whole document:

- Use common fonts and colors on all page types.
- Use common headers and footers on all page types.
- Use common page margins. (However, if it makes sense, feel free to combine sections that have only one text column with sections that have multiple text columns.)
- Align objects on all pages to a common design grid.

Standard page types

Typically, a user manual contains a subset of the following pages:

- a title page (front cover)
  See *Title page layout* 130.
- a page with legal information and version information (obligatory in some countries)
  See *Legal information layout* 133.
- a table of contents
  See *Table of contents layout* 137.
- separator pages that start each level 1 chapter
  See *Separator page layout* 143.
- body pages of chapters
  See *Body page layout* 146.
- an alphabetical index
  See *Index layout* 150.
- a back cover
  See *Back cover page and spine layout* 154.

> **Related rules**
>
> *Setting the type area* 59
>
> *Recommended screen layouts* 103

2.5.1   Title page layout

The title page is the "face" of your document.

Don't underestimate the importance of the title page design. You never get a second chance to make a first impression. If the title page gives readers the impression that your document is well-designed and professionally made, readers are likely to trust the document and will read it attentively.

While you don't necessarily need the help of a graphic artist for the design of the inner pages of a manual, consulting a media designer for the title page can be a good idea.

The title page is the key element to identify and recognize a document. It must enable readers to decide whether the document contains the information that they need.

Design tips

The title page must provide:

- the name and the manufacturer of the product
- if the product has versions: the version of the product
- the title of the manual
- if the title doesn't make this clear: the audience and goal of the manual

If your document will be translated: Bear in mind that texts may be longer in other languages.

Picture:

- If your product is hardware, include a photo or drawing of the product.
- If your product is software, include a logo of the product instead of a screenshot or a boxshot. A logo is more characteristic at first glance.

Basic design elements:

- A simple design only needs some background color and a few basic shapes.
- Other than on body pages, on a title page it's OK to add some purely decorative elements. However, make sure that decorative elements don't dominate the key information, in particular that they don't take visual precedence over the name of the product and the title of the manual.

- Sometimes you can make use of a key element taken from a product logo, or you can use another shape that's characteristic of the product. Later, maybe you can even use the same element or shape also within the document, for example, as a bullet for lists or for marking headings.

- For the main document title, inverted text (a light color on dark background) is often a good solution. Inverted text can put a heavy emphasis on the text without needing to be very large.

- Arrange all elements harmoniously (see *Use the golden ratio* 33).

Examples

To get started, you can adapt one of the following generic examples. Change colors, resize elements, and add pictures as required.

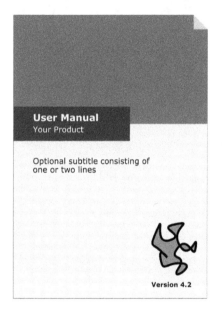

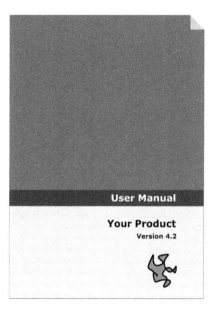

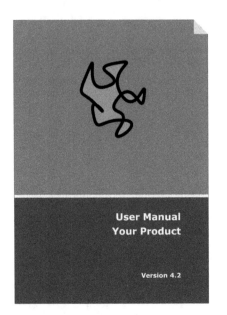
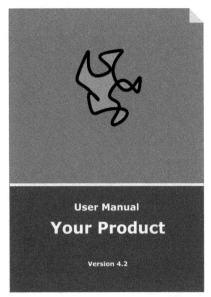

Related rules

Legal information page layout 135

Back cover page and spine layout 154

2.5.2   Legal information page layout

In many countries, printed documents must include a special page that indicates who's responsible for the document's contents. In addition, you need to provide contact information. Copyright notices and other legal information can also go here.

Design tips

Typically, the text is positioned near the bottom of an otherwise completely empty page (there's no other text, no header, no footer, and no page number).

The text is printed in a font size that's one or two points smaller than the body text of the document. The use of the small font signals to your readers that reading this page usually isn't required.

If you have right pages and left pages, put the legal information onto a left page. It's less prominent there than on a right page.

Example

ⓘ **Important:** Always check with your legal department or with an attorney which information you need to include and where you need to make this information available. If your legal department or attorney gives you any prefabricated materials to include, don't make any modifications to these materials without consultation.

The following picture shows a typical page with legal information.

YourProduct – User Manual
Edition <Month>, <Year>

YourCompany
YourStreet
YourCity and YourPostalCode

Tel.: 123–456789–0
Fax: 123–456789–1
Email: info@yourcompany.com
Internet: www.yourcompany.com

CEO, registration office, registration number, tax number
responsible for the content: first name, last name

(DE)

Gestaltungsvorschlag mit deutschem Text:

IhrProdukt – Benutzerhandbuch
Ausgabe <Monat>, <Jahr>

IhrUnternehmen
IhreStraße
IhrePostleitzahl IhreStadt

Tel.: 123–456789–0
Fax: 123–456789–1
E-Mail: info@ihrunternehmen.de
Internet: www.ihrunternehmen.de

Geschäftsführung, Handelsregister, Registernummer, Ust-Identifikationsnummer
Verantwortlich für den Inhalt: Vorname Nachname

Related rules

Title page layout 130

2.5.3 Table of contents page layout

The table of contents is a map of your document. Design the table of contents so that it helps your readers:

- to grasp the structure of the document at a glance
- to find a particular topic as quickly as possible

The key elements of a table of contents are:

- a heading that says that it's the table of contents
- a hierarchical list of all chapters within the document
- page numbers of the beginning of each chapter

Design tips

General recommendations:

- Always start the table of contents on a right page.
- Use the same basic page settings as for body pages (see *Body page layout*), such as page margins, headers, and footers. If you have a running header, have it say "Contents" on all pages of the table of contents. You can also omit a running header here because readers can easily identify the table of contents just by its appearance.
- Use only one text column. A two-column layout may only be appropriate if you also use a multi-column layout within the document or if you need to fit the whole table of contents onto a single page.
- If your document will be translated into foreign languages: Bear in mind that headings may be longer in other languages.
- Name the table of contents just *Contents*. Avoid the verbose form *Table of Contents*.
- Don't include an entry for the table of contents within the table of contents.

> (DE)
>
> Bevorzugen Sie als Überschrift im Deutschen die kurze Form *Inhalt*, nicht *Inhaltsverzeichnis*.

Visualizing structure:

- Clearly visualize the hierarchical structure of the document.
- Use different font sizes, but avoid using more than two.

- Use a font size and line spacing that allow a sufficient number of entries on one page. Readers should be able to grasp the document's structure immediately. For the lowest-level entries, use the same font size as for body text.

- Use space before and space after top-level headings to emphasize their importance.

- Make top-level headings bold.

- If you use a particular color for heading paragraph styles (see *Heading paragraph styles* 187), use this same color also in the table of contents.

- Optionally, you can add a fine line above or below top-level headings to emphasize the beginning of a new section.

- Make low-level headings gray or italic.

Chapter numbers:

If your headings include chapter numbers, also show these numbers in the table of contents. (However, in general it's better not to use chapter numbers at all; see *Heading paragraph styles* 187.)

Guides:

If your text column is wide and the headings are short, the reader's eyes must bridge a long distance between the end of a listed heading and the corresponding page number. Many readers slip into a wrong line then and end up on a wrong page. To prevent this from happening, consider adding a fine line or a series of dots as a visual guide.

Don't start the dots immediately after the last letter of the heading, but first add about 2 or 3 space characters. Add the same amount of space or even a little bit more between the end of the guide and the page number.

Example

The following picture shows a table of contents where the given design tips have been implemented.

Contents

5

Formatting page numbers

Orthodox typographers use the same character style for all page numbers because this looks more uniform than using the character style of the heading.

With the reader in mind, however, it's sometimes better to ignore this rule and to use the same style for the page number that you've used for the corresponding heading. This better visualizes the structure and minimizes the risk of slipping into the wrong line.

Compare the following variations:

- on the left side is the version that uses identical character styles for all page numbers
- on the right side is the version that uses different character styles for page numbers

You can add dots or thin lines as a visual guide between headings and page numbers. This is especially important if you use the same character style on all heading levels.

The following picture shows both versions of the table of contents with dots:

- on the left side is the version that uses identical character styles for all page numbers
- on the right side is the version that uses different character styles for page numbers

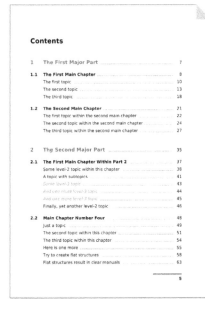

Options if a table of contents gets very long

If a table of contents gets so long that you're afraid that it might overwhelm readers, only include headings down to a particular level. For example, include heading levels 1 and 2 in the table of contents, but don't include heading levels 3 and 4.

This approach can be especially effective in reference manuals, where you often have a huge number of very short sections.

> **Important:** If you don't list all subchapters in the table of contents, make sure that users can easily find a particular subchapter anyway. For example, arrange subchapters that aren't included in the table of contents alphabetically or numerically. Also make sure that you have a good alphabetical index.

As an option, you can also add a mini table of contents (mini TOC) at the beginning of each parent chapter. A mini TOC is a small, partial table of contents on the first page of a major chapter and lists all subchapters within this chapter.

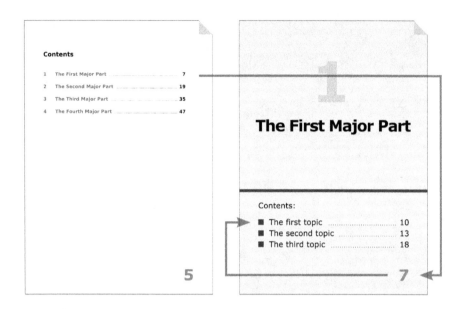

Contents

5

The First Major Part

Contents:

7

 Related rules

Index page layout 150

2.5.4   Separator page layout

Separator pages are optional. They mark the beginning of a new major part of a document.

Separator pages should be designed to stand out. Readers should be able to identify them clearly when thumbing through the document.

The key element of a separator page is white space or a large single-colored area. Either of these elements clearly sets apart the separator page from all other pages.

Design tips

General design:

- Make sure that separator pages are always right pages.
- Use a white background or a single-colored background. Use a light color rather than a dark color, especially if the document is shipped electronically so that users must print it on their office printers.
- Make the part number or part title large and visually prominent.
- When possible, align the elements on the separator page with elements on the title page.
- If your document will be translated: Bear in mind that texts may be longer in other languages.

Contents:

- Usually, don't include anything but the part number or part title.
- Optionally, you can add a brief description of the part's contents (maximum length: 3 sentences). If you add a description, design it in a way that's clearly different from standard body text.
- If you use part labels (see *Body page layout* 146), these labels should also be included on the separator page.

Examples

The following examples show some separator page layouts compared to the layouts of the title page. Note that in each design, there's at least one element that's repeated from the title page or that's aligned with an element on the title page.

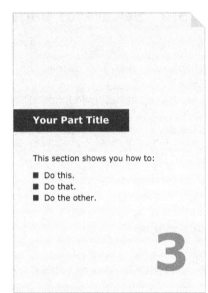

Your Part Title

This section shows you how to:

- Do this.
- Do that.
- Do the other.

3

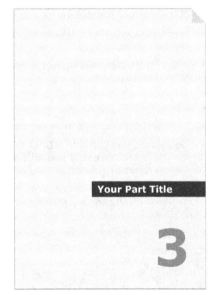

Your Part Title

3

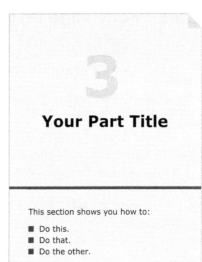

3

Your Part Title

This section shows you how to:

- Do this.
- Do that.
- Do the other.

2.5.5   Body page layout

Body pages are the standard pages that include the topics of your document.

The answer to the question of what's the optimum layout depends on the nature of your document and its content. In particular, it depends on how many pictures you'll include.

The key elements on body pages are:

- a running header that shows the title of the current chapter or the title of the parent chapter
- one or more text columns, or a margin column plus a text column
- the page number

Design tips

The design of the body pages is the core task of designing your document. It essentially involves *all* basic document design rules:

- *Layout basics* 15
- *Setting the type area* 59
- *Choosing fonts and spacing* 85
- For details on setting up styles, see *Recommended paragraph styles* 179 and *Recommended character styles* 219.

Example

The following picture shows a typical body page layout in relation to the title page layout.

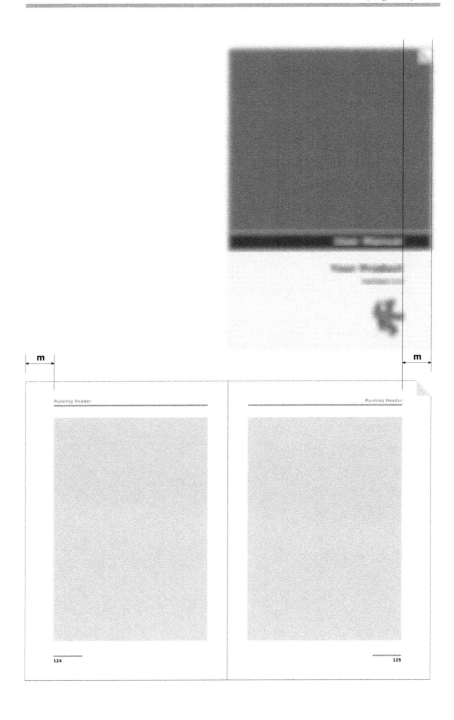

Part labels and section labels

If you want to use part labels or section labels as a navigation tool for your readers, you must add these labels to each right body page.

- On the label, use a dark background color so that the label is visible even when the book is almost closed. If your document is going to be printed in color, use a different color for each major section. If you want to keep your document from becoming too colorful, you can use different shades of the same color.

- If labels are dark, use white text.

- Arrange the labels from top to bottom. Start with the first section at the very top of the page, or start below the header at the top of the print area. With the last section, end at the very bottom of the page, or end above the footer at the bottom of the print area.

- Make the labels large enough so that there's enough space even for the longest label text. If your document is going to be translated into foreign languages, bear in mind that texts may be significantly longer in other languages.

- Check with your printer whether you need to extend the labels a few millimeters beyond the rim of the page (the so-called "bleed"). Depending on the printing process, this may be necessary to avoid a thin white line from appearing.

- Make all labels equally wide.

- When possible, don't overlap the labels.

- Make the labels as flat as possible so that they don't interfere with the pages' content.

Example:

Running Header

Running Header

Part 1
Your Part Title

124

125

2.5.6 Index page layout

An alphabetical index provides access to information via keywords. The design of the index must enable readers to find a particular keyword as quickly as possible.

The key elements of an index are:

- a heading that says that it's the index
- subheadings for all letters for which the index includes entries
- main entries
- subentries
- page numbers

Design tips

General page settings:

- Always start the index on a right page.
- Use the same basic page settings as for body pages, such as page margins, headers, and footers (see *Body page layout*). If you have a running header, have it say "Index" on all pages of the index. You can also omit a running header here because readers can easily identify the index just by its appearance.
- Within these limits, use a 2-column layout or a 3-column layout, regardless of how many columns your body pages have. Index entries typically are short, so with a 2-column or 3-column layout you can make your index significantly shorter and clearer.

 (DE)

Bevorzugen Sie als Überschrift im Deutschen die kurze Form *Index* oder *Stichwörter*, nicht *Stichwortverzeichnis*.

Subheadings for each letter:

- Add a subheading for each letter for which there are entries in the index.
- To make the beginning of a section visually more prominent, you can add a line above each letter. You can also use some color here, either for the letters or for the lines. To prevent your design from becoming too colorful, it's a good idea to use a color that already exists somewhere else, such as your corporate color, or the predominant color of your product logo.

(DE)

Behandeln Sie Umlaute nicht gesondert. Behandeln Sie *Ä* wie *A*, *Ö* wie *O* und *Ü* wie *U*.

- Fügen Sie also *keine* zusätzlichen Abschnitte für die Umlaute ein.
- Sortieren Sie ein *ä* so ein wie ein *a* (nicht wie *ae*), ein *ö* so wie ein *o* (nicht wie *oe)* und ein *ü* so wie ein *u* (nicht wie *ue*).

Behandeln Sie ein *ß* wie *ss*.

Entries, subentries, and page numbers:

- For the index entries, use the same font size as for body text, or up to 2 points smaller.
- To set off subentries from main entries, indent the subentries and make them italic.
- For the page numbers, use the same font and the same font style as for entries and subentries.
- If your authoring tool supports this feature: When a subentry "turns a page" (that is, when it's continued from a right-hand page to a left-hand page), repeat the main entry on the new page, followed by the word "(continued)" in parentheses.

 (DE)

Im Deutschen ist für „*(continued)*" der Zusatz „*(Fortsetzung)*" üblich.

Example

The following picture shows a typical 2-level index that you can easily adapt to fit your own layout.

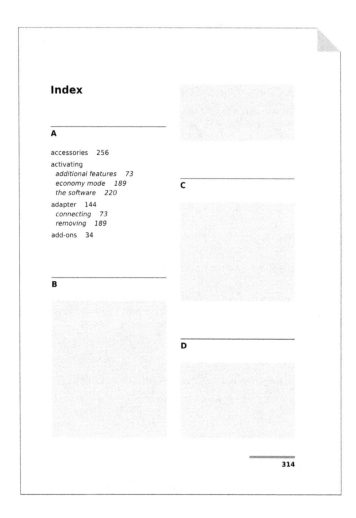

Index

A

accessories   256

activating
  additional features   73
  economy mode   189
  the software   220

adapter   144
  connecting   73
  removing   189

add-ons   34

B

C

D

314

Punctuation within index entries

Avoid using a comma before the first page reference. This clutters the layout.

Instead of a comma, add two or three space characters between the index entry and the first page number.

Add commas only between subsequent references. (Note: Usually, you shouldn't have any subsequent references in a well-written index. Instead, find index entries that are unique.)

✘ **No:**   *printing reports, 25, 57, 82*

✔ **Yes:**   *printing reports   25, 57, 82*

✔ **Top:** *printing reports*
         preparing   25
         procedure   57
         troubleshooting   82

 Related rules

Table of contents page layout 137

2.5.7 Back cover page and spine layout

The back cover is a good place to add identifying information, which is often more important for logistic purposes rather than for the readers. For example, you can add:

- the date of publication
- the version number
- the print number and language code
- the order number of the document
- the barcode

In addition, you can add contact information, such as a hotline number.

The information on the spine is much more important than the information on the back cover. The information on the spine helps readers to identify the document when it's standing on a bookshelf in a row with other books. The key information on the spine is:

- the identification of the product that the document describes (product name)
- the identification of the document (document title)

Design tips

Back cover:

- Design the back cover to be as plain and simple as possible. Use a lot of space. Don't position texts and barcodes somewhere in the middle of the page but near the borders.
- Print *internal* information vertically so that readers can easily distinguish it from the information that's relevant for them.
- When possible, use a simplified continuation of the front cover's design. Design the front cover, the spine, and the back cover as one unit.

Spine:

Unless your document has more than about 500 pages (not recommended for user manuals), you need to print the information on the spine vertically.

- In the U.S., the U.K., and Scandinavia, book spine titles are most often written with the title oriented top to bottom (as in the second and third of the following examples).
- In most of continental Europe, book spine titles are usually written from the

bottom to the top (as in the first of the following examples).

- Use the direction that's more common in the countries in which you mainly sell your product.

- When in doubt, use the direction that lets the title begin near the border of the spine rather than in the middle, depending on your particular design (compare examples below).

- If your document will be translated: Bear in mind that texts may be longer in other languages.

 (DE)

Im deutschen Sprachraum ist zu mehr als 90% die Anordnung von unten nach oben üblich.

Examples

The following pictures show some examples for your inspiration. Note how some common elements span the back cover page as well as the spine and the title page. All titles on the spine begin near the border.

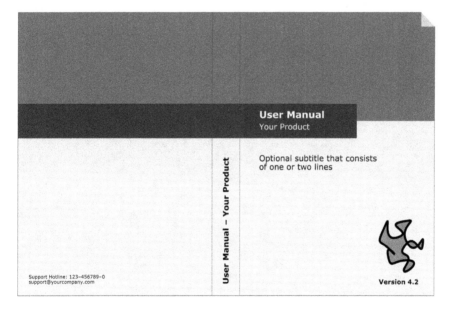

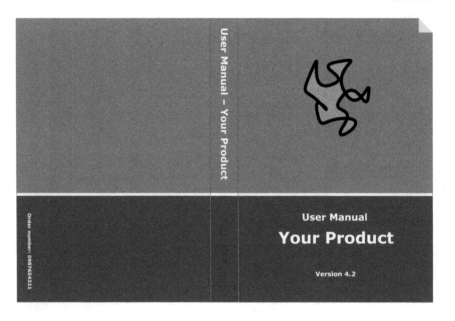

User Manual – Your Product

Order number: 098765
54321

User Manual

Your Product

Version 4.2

User Manual – Your Product

Order number: 0987654321

User Manual

Your Product

Version 4.2

Final page for electronically shipped documents

If you ship a document electronically, a spine doesn't exist and a traditional back cover page doesn't make sense. Alternatively, you can:

- End the document with the final page of the index.
- Add a page with contact information.

To set off a dedicated final page visually from the other pages of the document, keep it largely empty and place the contact information near the bottom.

Contact:

YourCompany
YourStreet
YourCity and YourPostalCode

Tel.: 123–456789–0
Fax: 123–456789–1
Email: info@yourcompany.com
Internet: www.yourcompany.com

Related rules

Title page layout 130

2.6   Recommended table designs

The purpose of tables can be:

- to structure information
- to arrange information
- to visualize relationships

Avoid anything that doesn't help achieve these purposes. In particular, use as few lines as possible to avoid unnecessary clutter (see also *Use clear and simple design* 19).

Types of tables

- In a printed document, all tables are static.
  See *Static tables design* 160.
- In online help, you can have static tables as well, but some authoring tools and JavaScript libraries also enable you to create dynamic tables.
  See *Dynamic tables design* 172.).

  Dynamic tables can:

  - be sortable
  - expand to show additional information
  - highlight particular information when readers hold the mouse pointer over a table cell

Related rules

Recommended paragraph styles 179

2.6.1   Static tables design

Tables structure information and visualize relationships.

Tables don't necessarily have to look like classical tables.

- Often, tables don't need lines.
- Often, tables don't need row headings and column headings.

In general, use as few lines as possible. This avoids unnecessary clutter and keeps your design clear and simple.

Basic tables with and without lines

Excessive lines don't make it easier to read a table, but instead make it harder. Do without lines whenever you can.

If you use light background colors and shades of gray instead of lines, your tables will look much more professional. You can still indicate "lines" by leaving some white space between the different areas.

If you don't want to do without lines completely, you should at least try to reduce the number of lines. For example, omit the lines at the sides of a table.

Make the remaining lines as thin as possible. As a rule of thumb, don't make dark lines thicker than the lines in the letter "I" of the used body font. Lines in light colors can be up to 5 times wider.

✖ **No:**

Component	Feature A EUR	Feature B EUR	Feature C EUR
Device A	25.045	134.765	254.112
Device B	7.234	37.765	−749
Device C	457.019	98.025	1.879

✘ No:

Component	Feature A EUR	Feature B EUR	Feature C EUR
Device A	25.045	134.765	254.112
Device B	7.234	37.765	–749
Device C	457.019	98.025	1.879

✔ Yes:

Component	Feature A EUR	Feature B EUR	Feature C EUR
Device A	25.045	134.765	254.112
Device B	7.234	37.765	–749
Device C	457.019	98.025	1.879

✔ Yes:

Component	Feature A EUR	Feature B EUR	Feature C EUR
Device A	25.045	134.765	254.112
Device B	7.234	37.765	–749
Device C	457.019	98.025	1.879

✔ **Yes:**

Component	Feature A EUR	Feature B EUR	Feature C EUR
Device A	25.045	134.765	254.112
Device B	7.234	37.765	−749
Device C	457.019	98.025	1.879

To help readers stay in the correct row, you can use zebra stripes instead of lines.

✔ **Yes:**

Option one	32 GB	12 ms	12.000 rpm	$ 249,–
Option two	128 GB	16 ms	17.000 rpm	$ 198,–
Option three	256 GB	8 ms	9.000 rpm	$ 359,–
Option four	8 GB	24 ms	21.000 rpm	$ 279,–

Layout tables

Tables don't always have to look like tables. Often, you can use a borderless, transparent table to position text and pictures.

In most authoring tools, dashed lines indicate a borderless table. When the document is printed, however, the table becomes invisible:

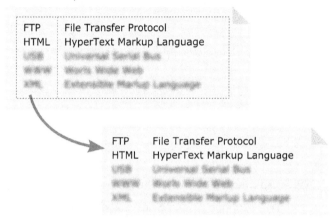

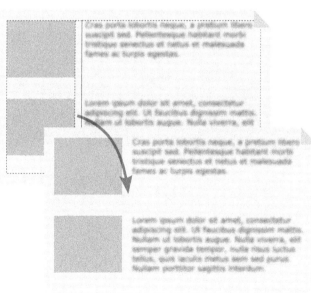

Multi-level headings

Often, you can simplify and shorten column headings or row headings by adding a main heading that spans several subheadings.

	Model		
Add-on	2000	3000	4000
Accessory A	$179,90	$189,90	$199,90
Accessory B	$1.059,90	$1.159,90	$1.259,90
Accessory C	$79,90	$89,90	$99,90

	Desktop models		Mobile models	
Add-on	2000–D	3000–D	2000–M	3000–M
Accessory A	$179,90	$189,90	$199,90	$209,90
Accessory B	$59,90	$159,90	$259,90	$359,90
Accessory C	$79,90	$89,90	$99,90	$109,90

Matrixes

To show interrelationships, you can design a table as a matrix.

Within the matrix, you can use text or symbols. If you use symbols, add a legend.

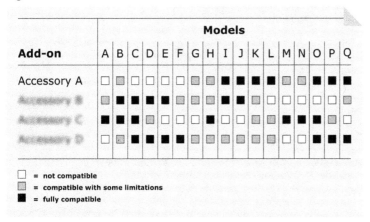

Add-on	Models																
	A	B	C	D	E	F	G	H	I	J	K	L	M	N	O	P	Q
Accessory A	□	▥	□	□	□	□	▥	▥	■	■	■	■	■	▥	▥	■	■
Accessory B	▥	■	■	■	■	▥	▥	▥	■	■	▥	□	□	□	□	□	▥
Accessory C	■	■	■	▥	□	□	□	■	□	□	▥	■	■	■	▥	□
Accessory D	□	▥	■	■	■	■	■	▥	▥	▥	▥	▥	▥	□	□	■	■

□ = not compatible
▥ = compatible with some limitations
■ = fully compatible

If a matrix is symmetric, you can simplify it by omitting vice-versa relationships.

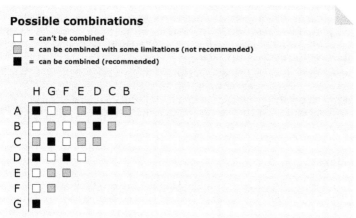

Possible combinations

□ = can't be combined
▥ = can be combined with some limitations (not recommended)
■ = can be combined (recommended)

	H	G	F	E	D	C	B
A	■	□	▥	▥	■	■	▥
B	□	▥	□	▥	■	▥	
C	▥	■	□	▥	▥		
D	■	□	■	□			
E	□	▥	▥				
F	□	▥					
G	■						

Alternative arrangements are:

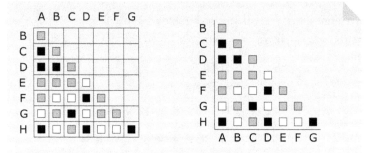

Which font size and font style?

Within the table cells, use the same font and the same font size as in the body text of your document. This has two important advantages:

- It looks consistent.
- You can use all body text paragraph styles and character styles also in your tables, for example, styles for lists, notes, warnings, and so on.

Don't use a font size within column headings or row headings that's larger than your body font size. Also, don't make text in column headings or row headings bold. The reason is that only important information should be large and bold. The important information, however, is in the body of the table, *not* in its head. If you want to set off the heading visually, use a different background color for the heading than for the body of the table.

When column headings are very long, you can even use a font size that's up to 2 points *smaller* than for the body of the table. In printed documents, using a smaller font size can be a very elegant solution and usually doesn't pose any problems. In online help, however, be careful that the font size is still legible.

Which alignment?

Use the same alignment for *all* cells within a column *and* for the column heading. For example, if you right-align the cells' contents, also right-align the column heading.

In general, use the same alignment for all columns of a table. The leftmost column, however, should always be left-aligned, so it can be an exception.

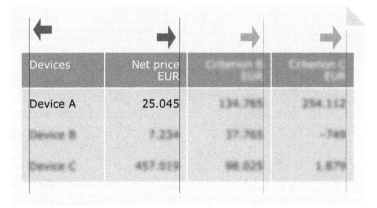

If table headings are boxed, center them vertically. If there are headings that have fewer lines of text than other headings, align the top of the short headings with the top of the longest heading.

If table headings are unboxed, align them at the bottom, near the start of the columns that they label.

Which cell padding?

As a rule of thumb, a good value for the distance between the border and the text is the width of the small letter "m" of the used font and font size. The bottom margin often needs to be slightly larger to provide a harmonic visual impression (see *Trust your visual judgment* 35).

Note that typically most table cells aren't completely filled with text, so most margins will be effectively larger than the configured padding.

What column widths?

To provide coherence, keep columns close together. Don't make them wider than necessary.

If your document will be translated: Bear in mind that texts may be longer in other languages.

Never space out the columns of a table merely to fill out the full page width. If the table is smaller, this is perfectly OK. Left-justify the table within the text column.

✖ No:

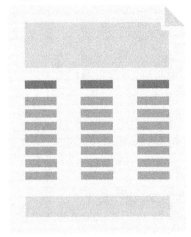

✔ **Yes:**

Sometimes, a column title is significantly longer than what goes into the cells of the column. Don't make a column overly wide only because of the length of its title. If you have a very long column title, do one of the following instead:

- Rephrase column titles and put them in note form.
- Use a multi-level heading if that makes sense.
- Run the text vertically or diagonally.
- Slightly decrease the character spacing.
- Use a font size up to 2 points smaller.
- Add line breaks within the column heading.

✖ **No:**

Devices	Column with a very long title	Other column
Device A	25.045	294.112
Device B	7.234	-749
Device C	457.019	1.879

✔ **Yes:**

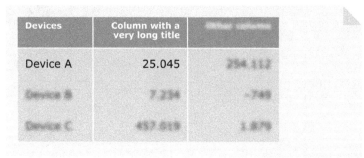

Devices	Column with a very long title	
Device A	25.045	254.112
Device B	7.234	~749
Device C	457.019	1.879

When you create online help, HTML makes it possible to set up auto-sizing tables. An auto-sizing table automatically adopts its width to the current width of the browser window. Within the table, either only one column can change its width, or all columns can change their widths proportionally.

Many tables have one column that clearly contains the bulk of the content. When that's the case, a good solution is to make only the width of just this column variable, but to make the widths of all other columns static. This is also the most robust solution if you create online help and printed documents from the same text base (single source publishing).

Should you use table titles?

If you can, avoid having table titles.

Table titles usually don't add any valuable information. If you have a well-designed document, it should be clear from your subheadings and from the column titles of your tables what each table is about.

Usually, you only need table titles if you want to add a table of tables at the end of your document. However, having a table of tables is dispensable or even obstructive in most cases, too. Having a table of tables only makes sense if users want to look up specific data that they can't easily find via the table of contents or via the index.

If you do decide to add table titles, add the titles above the tables rather than below. Other than with images (where figure titles usually go below the image) tables often span several pages. A table title below the table thus wouldn't be visible when a reader starts reading the table.

Should you number tables?

Only add table numbers if your content is going to be published as a printed document or PDF file and if you plan to generate a table of tables (usually not recommended).

When adding table numbers, don't number them throughout the document. Instead, number them chapter by chapter. This makes finding a particular table faster.

If you generate online help from the same text base as the printed document (single source publishing), make sure that the table numbers don't appear in the online version, where they don't make sense.

✘ **No:**   *Table 57. Examples and Demos*

✔ **Yes:**   **Table 3-2. Examples and Demos**

Should you allow page breaks within tables?

In general, avoid breaking tables, and especially avoid breaking small tables. With long tables, however, it's usually better to tolerate a page break within the table rather than to end up with half an empty page before a table on a new page.

Enable the following features if your authoring tool supports them:

- If you can't avoid breaking a table, keep at least two rows of the table on each page.
- If it isn't obvious what information each column of a table contains, repeat the column headings after a page break on the new page.
- If a table has a table title, repeat the table title on the new page, followed by the text "*(continued)*" in italic font style.

> ▬ (DE)
>
> Im Deutschen ist für „*(continued)*" der Zusatz „*(Fortsetzung)*" üblich.

Related rules

Dynamic tables design 172

2.6.2 Dynamic tables design

In online help, dynamic (interactive) tables can provide some additional value compared to static tables on paper.

Some help authoring tools come with integrated support to create dynamic tables. In all other cases, the easiest way to implement dynamic tables is by using one of the major (open source) JavaScript libraries, such as jQuery or MooTools.

Sortable tables

In a sortable table, when users click a column heading, the table is sorted alphabetically or numerically according to this column.

Provide some visual clue that column headings are clickable. Small triangles or arrow symbols are the most commonly used indicators. The advantages of triangles and arrows are:

- Triangles and arrows can indicate the current sort order.
- You can provide one triangle or arrow to sort in ascending order, and one to sort in descending order.

In addition to the triangles or arrows, you can also use color to highlight the column that currently determines the sort order, or you can make the column heading bold.

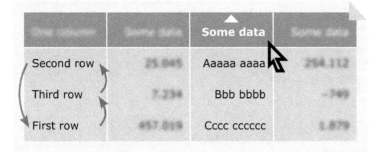

Tables with expandable areas

In a table with expandable areas, only that part of the information is visible by default that's important for all readers. Details that concern only a minority of readers are hidden. When readers click an expandable row or column, the table dynamically expands. When they click again, the table shrinks back to its original size.

You can design a table so that only parts of a row or parts of a column can be hidden, or you can design the table so that complete rows or complete columns can be hidden.

Always provide a visual clue that a row or column can be expanded or collapsed. Triangle symbols are commonly used for this purpose.

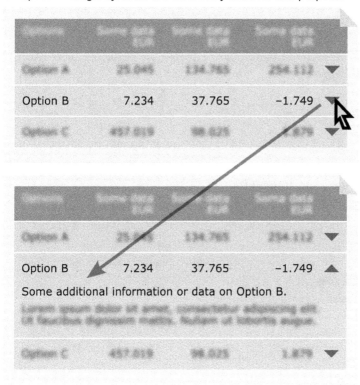

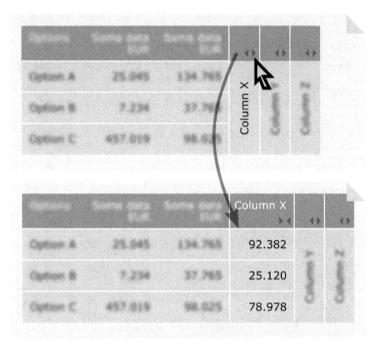

Tables with dynamic highlighting

Dynamic highlighting effects are reading aids that are especially helpful in large tables. The highlighting can prevent the readers' eyes from losing track of the current row and column.

When readers hold the mouse pointer over a table cell, you can highlight:

- single table cells
- complete rows or complete columns
- relationships

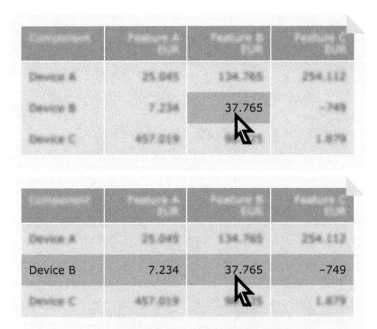

Component	Feature A EUR	Feature B EUR	Feature C EUR
Device A	25.045	134.765	254.112
Device B	7.234	37.765	–749
Device C	457.019	98.125	1.879

Component	Feature A EUR	Feature B EUR	Feature C EUR
Device A	25.045	134.765	254.112
Device B	7.234	37.765	–749
Device C	457.019	98.125	1.879

Component	Feature A EUR	Feature B EUR	Feature C EUR
Device A	25.045	134.765	254.112
Device B	7.234	37.765	–749
Device C	457.019	98.125	1.879

Component	Feature A EUR	Feature B EUR	Feature C EUR
Device A	25.045	134.765	254.112
Device B	7.234	37.765	–749
Device C	457.019	98.125	1.879

To achieve the highlighting effect, you can use:

- a distinction by color
- a distinction by size
- a combination of both color and size

Component	Feature A EUR	Feature B EUR	Feature C EUR
Device A	25.045	134.65	254.112
Device B	7.234	37.765	-749
Device C	457.019	98.025	1.879

Component	Feature A EUR	Feature B EUR	Feature C EUR
Device A	25.045	134.65	254.112
Device B	7.234	37.765	-749
Device C	457.019	98.025	1.879

Component	Feature A EUR	Feature B EUR	Feature C EUR
Device A	25.045	134.65	254.112
Device B	7.234	37.765	-749
Device C	457.019	98.025	1.879

Related rules

> *Static tables design* 160

2.7 Recommended paragraph styles

Paragraph styles determine the base typeface that's used in a paragraph, line spacing, indentations, borders, and background colors.

Professional authoring tools also let you control automatic hyphenation and page breaks in various ways.

> ⓘ **Important:** Not all authoring tools support all settings. If your authoring tool doesn't support a particular setting that we recommend, simply ignore this setting.

General recommendations

Provide as few paragraph styles as possible. A clearly written document doesn't need many different styles.

Create styles only for tagging particular types of information (see *Create styles semantically* 47).

For all paragraph styles:

- When configuring paragraph properties, add space *before* paragraphs rather than *after* paragraphs.

  If you mix space before and space after paragraphs, some authoring tools only use the larger space and ignore the smaller space. Other authoring tools sum up both spaces. In either case, the total distance between two paragraphs is hard to control due to the large number of possible combinations.

  In practice, it's not always possible to avoid having space after paragraphs in all cases, but at least try to avoid it if you can. This may save you a lot of trouble.

- Don't ignore the settings for orphan lines and widow lines.
  If your authoring tool supports it, set orphan control and widow control to a value of 2 for all paragraph styles.

  Orphan control specifies the minimum number of lines in a paragraph before a page break. If the number of lines at the end of the page is less than the specified number of lines, the paragraph is automatically shifted to the next page.

  Widow control specifies the minimum number of lines in a paragraph after a page break. If the number of lines at the beginning of the new page is less than the specified number of lines, the position of the break is automatically adjusted.

━━ (DE)

Widow und *Orphan* haben im Deutschen keine wörtliche Entsprechung.

- Der analoge deutsche Begriff zu *Widow* (wörtlich „Witwe") lautet *Hurenkind* (letzte Zeile eines Absatzes erscheint als erste Zeile auf einer neuen Seite oder in einer neuen Spalte).

- Der analoge deutsche Begriff zu *Orphan* (wörtlich „Waisenkind") lautet *Schusterjunge* (erste Zeile eines Absatzes erscheint als letzte Zeile auf einer Seite).

Overview of recommended styles

The basic paragraph styles needed for every user assistance document are:

Purpose	Suggested style names	Suggested keyboard shortcuts
Headings:		
heading level 1	h**1**	[Alt]+[Shift]+[**1**]
heading level 2	h**2**	[Alt]+[Shift]+[**2**]
heading level 3	h**3**	[Alt]+[Shift]+[**3**]
For details, see: *Heading paragraph styles* 187.		
Subheadings:		
subheading that introduces a static section (always visible)	ss_s**u**bhead_static	[Alt]+[Shift]+[**u**]
subheading that introduces an expandable section (toggle)	se_subhead_**e**xpandable	[Alt]+[Shift]+[**e**]
For details, see: *Subheading paragraph styles* 194.		
Body text:		
standard paragraph	bo_b**o**dy	[Alt]+[Shift]+[**o**]
For details, see: *Body text paragraph style* 199.		
Procedures:		
paragraph that introduces a procedure	pi_**p**rocedure_intro	[Alt]+[Shift]+[**p**]
step of a procedure (with number)	ps_procedure_**s**tep	[Alt]+[Shift]+[**s**]

indented first-level paragraph in a procedure (without number)	pp_procedure_plain	–
For details, see: *Procedure paragraph styles* 201		
Lists:		
paragraph that introduces a list	li_**l**ist_intro	[Alt]+[Shift]+[**l**]
first-level item of a list (with bullet)	l1_list_level1_**b**ullet	[Alt]+[Shift]+[**b**]
indented first-level paragraph in a list (without bullet)	l1p_list_level1_plain	–
second-level item of a list (with bullet)	l2_list_level2_bullet	–
indented second-level paragraph in a list (without bullet)	l2p_list_level2_plain	–
For details, see: *List paragraph styles* 207		
Annotations: **Tips**	at_**t**ip	[Alt]+[Shift]+[**t**]
For details, see: *Note and warning paragraph styles* 213		
Annotations: **Notes**		
standard note	an_**n**ote	[Alt]+[Shift]+[**n**]
important note	ai_**i**mportant	[Alt]+[Shift]+[**i**]
For details, see: *Note and warning paragraph styles* 213		
Annotations: **Warnings**		
caution	ac_**c**aution	[Alt]+[Shift]+[**c**]
warning	aw_**w**arning	[Alt]+[Shift]+[**w**]
danger	ad_**d**anger	[Alt]+[Shift]+[**d**]
For details, see: *Note and warning paragraph styles* 213		

You can organize the styles hierarchically as follows:

Custom_Paragraph_Styles
- **Headings**
  - h1
  - h2
  - h3
- **Subheadings**
  - ss_subhead_static
  - se_subhead_expandable
- b0_body
- **Procedures**
  - pi_procedure_intro
  - ps_procedure_step
  - pp_procedure_plain
- **Lists**
  - li_list_intro
  - l1_list_level1_bullet
  - l1p_list_level1_plain
  - l2_list_level2_bullet
  - l2p_list_level2_plain
- **Annotations**
  - at_tip
  - **Notes**
    - an_note
    - ai_important
  - **Warnings**
    - ac_caution
    - aw_warning
    - ad_danger

(DE)

Wenn Sie lieber mit deutschen Bezeichnungen arbeiten, können Sie z. B. die folgenden Formatnamen und Tastenkürzel verwenden:

Zweck	Vorschlag für Formatnamen	Vorschlag für Tastenkürzel
Überschriften:		
Überschrift Ebene 1	ue1	[Alt]+[Shift]+[1]
Überschrift Ebene 2	ue2	[Alt]+[Shift]+[2]
Überschrift Ebene 3	ue3	[Alt]+[Shift]+[3]

Details siehe:
Heading paragraph styles 187.

Zwischenüberschriften:

Zwischenüberschrift, die einen statischen Abschnitt einleitet (immer sichtbar)	zs_zue_statisch	[Alt]+[Shift]+[z]
Zwischenüberschrift, die einen expandierbaren Abschnitt einleitet (Toggle)	ze_zue_expandierbar	[Alt]+[Shift]+[e]

Details siehe:
Subheading paragraph styles 194.

Grundtext:

Standardabsatz	tx_textkoerper	[Alt]+[Shift]+[x]

Details siehe:
Body text paragraph style 199.

Handlungsanweisungen:

Absatz, der eine schrittweise Handlungsanweisung einleitet	si_schritt_intro	[Alt]+[Shift]+[s]
Schritt einer schrittweisen Handlungsanweisung (mit Nummer)	sn_schritt_nummer	[Alt]+[Shift]+[n]
Eingerückter Fortsetzungsabsatz erster Ebene (ohne Nummer)	sf_schritt_folge	–

Details siehe:
Procedure paragraph styles 201.

Listen:

Absatz, der eine Liste einleitet	li_liste_intro	[Alt]+[Shift]+[l]
Listenpunkt erster Ebene einer Liste (mit Aufzählungszeichen)	l1_liste_ebene1_az	[Alt]+[Shift]+[a]
Eingerückter Folgeabsatz erster Ebene (ohne Aufzählungszeichen)	l1f_liste_ebene1_folge	–
Listenpunkt zweiter Ebene einer Liste (mit Aufzählungszeichen)	l2_liste_ebene2_az	–
Eingerückter Folgeabsatz zweiter Ebene (ohne	l2f_liste_ebene2_folge	–

Recommended paragraph styles

Aufzählungszeichen) Details siehe: *List paragraph styles* [207]		
Anmerkungen: **Tipps** Details siehe: *Note and warning paragraph styles* [213]	at_**t**ipp	[Alt]+[Shift]+[**t**]
Anmerkungen: **Hinweise** Standardhinweis Wichtiger Hinweis Details siehe: *Note and warning paragraph styles* [213]	ah_**h**inweis aw_**w**ichtig	[Alt]+[Shift]+[**h**] [Alt]+[Shift]+[**w**]
Anmerkungen: **Warnhinweise** Vorsicht Warnung Gefahr Details siehe: *Note and warning paragraph styles* [213]	av_**v**orsicht aw_**w**arnung ag_**g**efahr	[Alt]+[Shift]+[**v**] [Alt]+[Shift]+[**w**] [Alt]+[Shift]+[**g**]

Für die hierarchische Organisation gilt dann:

Eigene_Absatzformate
- **Überschriften**
  - ue1
  - ue2
  - ue3
- **Zwischenüberschriften**
  - zs_zue_statisch
  - ze_zue_expandierbar
- b0_textkoerper
- **Handlungsanweisungen**
  - si_schritt_intro
  - sn_schritt_nummer
  - sf_schritt_folge
- **Listen**
  - li_liste_intro
  - l1_liste_ebene1_az
  - l1f_liste_ebene1_folge
  - l2_liste_ebene2_az
  - l2f_liste_ebene2_folge
- **Anmerkungen**
  - at_tipp
  - **Hinweise**
    - ah_hinweis
    - aw_wichtig
  - **Warnhinweise**
    - av_vorsicht
    - aw_warnung
    - ag_gefahr

Related rules

Automate line breaks and page breaks 4↑

Create styles semantically 47↑

Create styles hierarchically 56↑

Use color with care 2↑

Think ahead about printing 36↑

Choosing fonts and spacing 85↑

Recommended table styles 159↑

Recommended character styles 219

2.7.1 Heading paragraph styles

The design of headings must emphasize the various functions that headings have:

- Headings mark the beginning of a topic.

- When readers skim a document, headings quickly communicate what the topic is about. Headings aid the decision whether or not to read the topic.

- In a printed document, headings communicate the hierarchy level of a topic.

  In online help, there is essentially no hierarchy (online help is hypertext even if you have a table of contents). For this reason, in online help, all headings have the same style.

Usually, you shouldn't have more than 3 hierarchy levels, so you need the styles:

- **h1**
  heading level 1

- **h2**
  heading level 2

- **h3**
  heading level 3

 (DE)

Namensvorschläge im Deutschen: **ue1**, **ue2**, **ue3**

This is a Level 1 Heading] ← h1

This is a Level 2 Heading] ← h2

This is a Level 3 Heading] ← h3

Basic settings

Always left align headings. Don't justify headings even if your body text is justified (not recommended).

In general, use the same font as for all other styles. If the used font is very wide, you can optionally use a related, narrower font (see *Which font?* 86). In most cases, however, it's enough to slightly reduce the character spacing as part of the character settings.

Don't use small caps because small caps significantly slow readers down.

Visual weight

In online help, use the same heading style for *all* topics, regardless of where a topic stands in the hierarchy of the table of contents.

In a printed document, increase the visual weight depending on the heading's hierarchy level. Make *h1* more prominent than *h2*, and make *h2* more prominent than *h3*.

To increase the visual weight, you can:

- use a larger font size
- use bold font style
- use color instead of black or gray
- add space above the paragraph

Spacing

Add more space before the heading than after the heading. Don't make the space after the heading larger than one line of text so that it's clearly visible that the following text belongs to the heading.

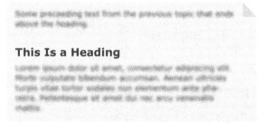

If you have a small type area, you can't avoid having headings that run across two lines. Make sure that the line spacing is visually smaller than the space below the heading. Bear in mind that the larger the font is, the smaller the relative line spacing should be (see *What line spacing?* 98).

Headings in narrow columns may take several lines

Lorem ipsum dolor sit amet, consectetur adipiscing elit. Ut faucibus dignissim mattis. Nullam ut lobortis augue. Nulla viverra, elit semper gravida tempor, nulla risus luctus tellus, quis iaculis metus sem sed purus. Nullam porttitor sagittis interdum.

Lines and background color

If you want to separate the distinct sections of your document very clearly, you can add lines or background colors to headings.

Some preceding text from the previous topic that ends above the heading.

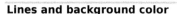

Headings structure your text

Lorem ipsum dolor sit amet, consectetur adipiscing elit. Morbi vulputate bibendum accumsan. Aenean ultricies turpis vitae tortur sodales non elementum ante placerat. Pellentesque sit amet dui nec arcu venenatis mattis.

Lorem ipsum dolor sit amet, consectetur adipiscing elit. Ut faucibus dignissim mattis. Nullam ut lobortis augue. Nulla viverra, elit semper gravida tempor, nulla risus luctus tellus, quis iaculis metus sem sed purus. Nullam porttitor sagittis interdum.

Cras porta lobortis neque, a pretium libero suscipit sed. Pellentesque habitant morbi tristique senectus et netus et malesuada fames ac turpis egestas.

Sed orci lacus, congue sit amet elementum ac, commodo et urna. Suspendisse bibendum ullamcorper tellus, quis egestas nunc laoreet vitae.

Another section

Praesent pellentesque eleifend tellus, sit amet suscipit lorem luctus eu. Integer sodales lobortis faucibus. Aliquam rhoncus quam sed magna cursus gravida. Suspendisse justo sapien, varius id iaculis ac, ornare at ligula.

One section

Donec libero sapien, gravida id interdum ut, egestas at erat. Sed non turpis erat.

Font color

If you want to bring some color into your document, an unobtrusive way of doing so is to give level 1 headings (*h1*) your corporate color or the predominant color of your product logo.

Note:
If you use color here, also use the same color in the table of contents (see *Table of contents page layout* 137).

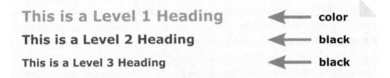

Should you add numbers?

In online help, heading numbers don't make any sense at all. In hypertext, essentially there's no given hierarchy and sequence—even if you have a table of contents as a navigation tool.

In printed manuals, chapter numbers are often included:

- because the authoring tool can do it automatically

- because chapter numbers are common in scientific textbooks, and these books are what developers and engineers are used to reading

Neither point is a good reason to use heading numbers. In user assistance, chapter numbers usually have more disadvantages than advantages.

The advantages of having chapter numbers are:

- Chapter numbers indicate the headings' hierarchy level. (However, if you have well-designed heading paragraph styles, this extra visualization shouldn't be necessary.)

- Support staff often prefer referring callers to chapter numbers rather than to page numbers because chapter numbers change less often when a document is updated.

The disadvantages of having chapter numbers are:

- Chapter numbers add clutter to the page without adding valuable information. In particular, chapter numbers clutter the table of contents, which makes it more difficult both to grasp the overall structure and to find a specific topic.

- Chapter numbers create the look and feel of a scientific paper, of a complicated textbook, or of legal text, instead of a document that's meant to encourage reading.

- In many authoring tools, numbered headings are significantly harder to handle and more error-prone than headings without numbers.

✘ No: *3.2.4.1 How to Write Clearly*
✔ Yes: *How to Write Clearly*

Tip:
Even if you do decide to number headings, you don't have to number the headings of all hierarchy levels. For example, you can number heading levels 1 and 2 but not number heading levels 3 and 4. This can help to make your table of contents much clearer. At least it avoids the longest, most difficult-to-read numbers.

If you number headings, don't include a period after the last number.

✘ No: *7.5.3. The Heading*
✔ Yes: *7.5.3 The Heading*

Don't use Roman numerals. Many readers don't know how to read them. Also, Roman numerals are longer than Arabic numerals.

✘ No: *IV The Heading*
✔ Yes: *4 The Heading*

✘ No: *Part III*
        1 The Heading
✔ Yes: *Part 3*
        1 The Heading

        or:

        Part C
        1 The Heading

If you number headings, format the numbers less prominently than the heading text. For example, make the numbers gray and less bold than the heading.

2.3.4 Headings structure your text

2.3.4 Headings structure your text

Headings structure your text

When should a heading start on a new page?

If you add automatic page breaks before headings so that each topic starts on a new page, this can save you a lot of manual work when fine-tuning the final formatting of your document.

The downside of beginning a new page for each topic is that it can make your document longer and adds more unused white space at the end of topics (for details, see *Automate line breaks and page breaks* 41).

As rules of thumb:

- If your document is shipped electronically and you don't expect most readers to print it, start *all* headings on a new page.
- If your document is shipped on paper, or if many readers will print it:
  - Start at least headings h1 und h2 on a new page.
  - Start all other heading levels on a new page only if the majority of topics isn't shorter than half a page. Avoid having an average of more than 30% of white space on your pages.

Special settings

h1

- Set to keep with the next paragraph on the same page.
- Set to keep all paragraph lines together on the same page.
- Automatically add a page break before the paragraph.

  If you have a layout with right pages and left pages, set up the style so that it always starts on a right page (that is on a page with an odd page number)
  .

h2

- Set to keep with the next paragraph on the same page.
- Set to keep all paragraph lines together on the same page.
- Optional: Automatically add a page break before the paragraph.

h3

- Set to keep with the next paragraph on the same page.
- Set to keep all paragraph lines together on the same page.
- Optional: Automatically add a page break before the paragraph.

Related rules

Subheading paragraph styles 194

2.7.2   Subheading paragraph styles

Subheadings (labels) communicate what a particular section within a chapter or within a help topic is about.

Subheadings are landmarks that help readers to skim a text for specific information and to decide beforehand whether or not to invest any time in reading.

In online help, subheadings often appear as expandable sections (toggles).

Subheadings should be:

- easily recognizable
- easily readable

Needed styles:

- **ss_subhead_static**
  subheading of static section

- **se_subhead_expandable**
  subheading of expandable section (only needed for online help)

 (DE)

Namensvorschläge im Deutschen:

- **zs_zue_statisch**
- **ze_zue_expandierbar**

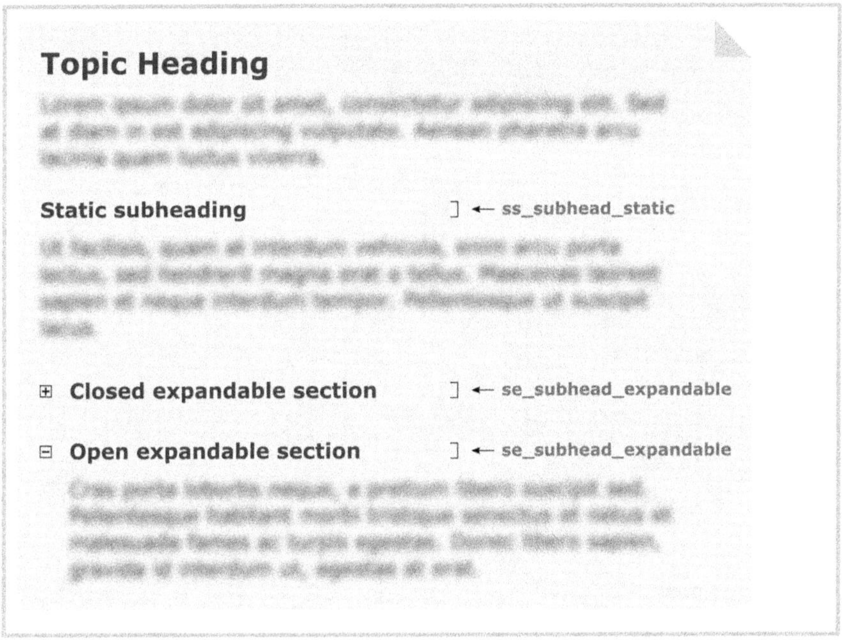

Topic Heading

(placeholder body text)

Static subheading   ] ← ss_subhead_static

(placeholder body text)

⊞ **Closed expandable section**   ] ← se_subhead_expandable

⊟ **Open expandable section**   ] ← se_subhead_expandable

(placeholder body text)

Basic design

The font size of subheadings doesn't have to be larger than the font size of body text. It's sufficient to make the text bold and to add some extra space before the paragraph.

If you decide to use a larger font size, don't make it more than 2 points larger than body text.

Always add more space *before* the subheading than after the subheading so that it's clearly visible that the following text belongs to the subheading. A good value for the distance between the preceding text and the subheading is about one blank line. The space after the subheading should be about one-third of this.

Clearly distinguish subheadings from major topic headings. One good way of doing so is to color subheadings gray. This makes them unobtrusive even if they're bold.

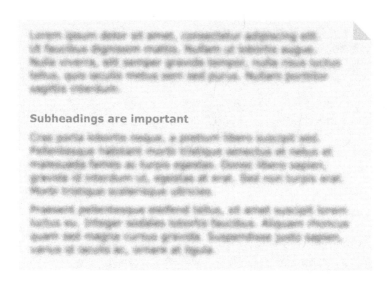

Subheadings are important

Design options

If you need to save space, you can put the subheadings at the beginning of the next paragraph. In this case, make sure that the subheading is clearly set off visually from the body text. Make the subheading bold, use some color, and add a few space characters after the subheading.

Tip:
Use your corporate color or the predominant color of your product logo.

Subheadings are important Cras porta lobortis neque, a pretium libero suscipit sed. Pellentesque habitant morbi tristique senectus et netus et malesuada fames ac turpis egestas. Donec libero sapien, gravida id interdum ut, egestas at erat. Sed non turpis erat.

In layouts that have a margin column, subheadings are typically printed within the margin column.

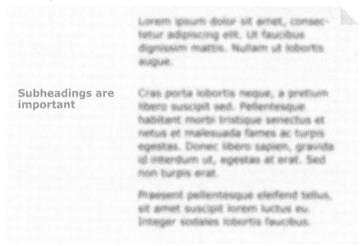

Design of expandable sections

Similar to static subheadings, labels of expandable sections should be easily recognizable and easily readable.

Add a symbol in front of the text. Plus/minus symbols or triangles are commonly used for this purpose. Don't invent any other symbols.

Don't underline expandable sections. Underlined text is commonly reserved for links that take readers to another topic. This is *not* the case here.

⊞ **Closed expandable section**

⊟ **Open expandable section**

Cras porta lobortis neque, a pretium libero suscipit sed.
Pellentesque habitant morbi tristique senectus et netus et
malesuada fames ac turpis egestas. Donec libero sapien,
gravida id interdum ut, egestas at erat.

▸ **Closed expandable section**

▾ **Open expandable section**

Cras porta lobortis neque, a pretium libero suscipit sed.
Pellentesque habitant morbi tristique senectus et netus et
malesuada fames ac turpis egestas. Donec libero sapien,
gravida id interdum ut, egestas at erat.

Special settings

ss_subhead_static

- Set to keep with the next paragraph on the same page.
- Set to keep all paragraph lines together on the same page.

se_subhead_expandable

- Set to keep with the next paragraph on the same page. (This has no effect in online help. It's only relevant if you generate a printed manual and online help from the same text base.)
- Set to keep all paragraph lines together on the same page. (This has no effect in online help. It's only relevant if you generate a printed manual and online help from the same text base.)

Related rules

Heading paragraph styles 187

2.7.3 Body text paragraph style

Body text is the standard text of your document. All text that has no special formatting is body text.

Needed style:

b0_body
Body text (standard paragraph)

 (DE)

Namensvorschlag im Deutschen: **t0_textkoerper**

Lorem ipsum dolor sit amet, consectetur adipiscing elit. Ut faucibus dignissim mattis. Nullam ut lobortis augue. Nulla viverra, elit semper gravida tempor, nulla risus luctus tellus, quis iaculis metus sem sed purus. Nullam porttitor sagittis interdum.   ← b0_body

Cras porta lobortis neque, a pretium libero suscipit sed. Pellentesque habitant morbi tristique senectus et netus et malesuada fames ac turpis egestas. Donec libero sapien, gravida id interdum ut, egestas at erat. Sed non turpis erat. Morbi tristique scelerisque ultricies.   ← b0_body

Space between paragraphs

Add enough space before the paragraph so that it's clearly visible where a new paragraph starts. Make the space larger than the space between two lines, but make it smaller than a blank line. A good value for the space before a body paragraph is about 1/4 to 1/2 the size of a blank line. There should still be enough cohesion so that it's visible which lines belong together.

First-line indent (optional)

In addition to the space between paragraphs, you can mark the beginning of each paragraph by setting up a small indent on the first line.

A good value for the indent is the value of the line height. Don't make the indent too large because then the text loses its cohesion and becomes more difficult to read.

Indents are common in printed documents, but they're rarely used in online

199

help.

Lorem ipsum dolor sit amet, consectetur adipiscing elit. Ut faucibus dignissim mattis. Nullam ut lobortis augue. Nulla viverra, elit semper gravida tempor, nulla risus luctus tellus, quis iaculis metus sem sed purus. Nullam porttitor sagittis interdum.

Cras porta lobortis neque, a pretium libero suscipit sed. Pellentesque habitant morbi tristique senectus et netus et malesuada fames ac turpis egestas. Donec libero sapien, gravida id interdum ut, egestas at erat. Sed non turpis erat. Morbi tristique scelerisque ultricies.

2.7.4 Procedure paragraph styles

A procedure is a description of the steps that users have to take to complete a task.

A procedure typically begins with an introductory phrase that states the goal of the procedure.

The introduction is followed by a series of numbered steps.

Needed styles:

- **pi_procedure_intro**
  introduces a procedure

- **ps_procedure_step**
  step of a procedure (with a number)

- **pp_procedure_plain**
  indented first-level paragraph within a procedure (without a number)

If you need to add a list within a procedure, use the list paragraph styles **l2_list_level2_bullet** and **l2p_list_level2_plain** (see *List paragraph styles* 207).

▬ (DE)

Namensvorschläge im Deutschen: **si_schritt_intro, sn_schritt_nummer, sf_schritt_folge**

This paragraph introduces the procedure: ⌋ ← pi_procedure_intro

1. The first step. ⌋ ← ps_procedure_step

2. The second step.
   You can use manual line breaks. ⌋ ← ps_procedure_step

3. The third step, introducing a list: ⌋ ← ps_procedure_step

   — This is a level 2 item. ⌋ ← l2_list_level2_bullet

   — One more level 2 item. ⌋ ← l2_list_level2_bullet

   This paragraph follows a list 2 item. ⌋ ← l2_list_level2_plain

4. The fourth step. ⌋ ← ps_procedure_step

   This is an indented plain paragraph. ⌋ ← pp_procedure_plain

Numbers

For the steps, use Arabic numerals. Don't use Roman numerals because many readers don't know how to read them. Also don't use letters instead of numerals.

Format the numbers so that they clearly stand out from the text. Use a hanging indent. In addition, you can make the numbers bold or use color. Procedures should clearly be recognizable as such when readers skim the text.

Tip:
If you use color, use your corporate color or the predominant color of your product logo.

This paragraph introduces the procedure:

1. The first step.
2. The second step.
   You can use manual line breaks within a step.
3. The third step.

If you prefer simplistic design, it's OK to leave out the periods after the numerals if the numerals clearly stand out from the text.

This paragraph introduces the procedure:

1 The first step.
2 The second step.
   You can use manual line breaks within a step.
3 The third step.

Division lines (optional)

If you want to emphasize a strong separation between the individual steps, you can add lines between the steps of the procedure.

This is adequate, especially for overall procedures or if performing each step takes a considerable amount of time.

Division lines are also a good method in a very short document (such as a small leaflet) where procedures are in close proximity to other information types.

This paragraph introduces the procedure:

1. The first step.

2. The second step.
And some text that follows.

3. The third step.

4. The fourth step.
And some text that follows.

Single-step procedures

If a procedure consists of only one step, don't add a number. Instead of adding a single "1," use another symbol, such as an arrow. Alternatively, format the step as regular body text.

This paragraph introduces the procedure:

→ This is the only step.

Indentation

Set up a hanging indent.

Tip:
Use the same indent that you use for lists (see *List paragraph styles* 207). You can then apply level 2 lists also within procedures. In addition, consistent indents considerably add to the professional look of your template.

✘ No:

✘ No:

✔ Yes:

If space is tight, you can alternatively use a run-in numbered list. A run-in list is less clearly laid out than a list with a hanging indent, but it's still much clearer than standard body text.

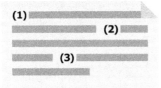

If a picture, a table, a warning, or another element relates to a particular step, indent this element along with the step. (This is usually only worth the effort if your authoring tool supports this feature without having to define a separate, indented version of each style.)

✘ No:

✔ Yes:

Special settings

pi_procedure_intro:

- Add some extra space before the paragraph so that the space before the procedure intro is slightly larger than the space before a body paragraph. This sets the procedure clearly apart from the rest of the text.

- Set to keep with the next paragraph on the same page.

ps_procedure_step:

- Set to keep all paragraph lines together on the same page.

pp_procedure_plain:

- Set to keep with the previous paragraph on the same page.
- Set to keep all paragraph lines together on the same page.
- Add less space before this paragraph than before *ps_procedure_step*.

Related rules

List paragraph styles 207

2.7.5 List paragraph styles

Lists break down information into small pieces and present these pieces of information in a clear, well-structured way.

There's no strict sequence in the items in the list, so list items are *not* numbered. Typically, small bullets are used to mark the beginning of each list item.

Needed styles:

- **li_list_intro**
  introduces a list

- **l1_list_level1_bullet**
  first-level list item (with bullet)

- **l1p_list_level1_plain**
  indented first-level paragraph within a list (without bullet)

- **l2_list_level2_bullet**
  second-level list item (with bullet)

- **l2p_list_level2_plain**
  indented second-level paragraph within a list (without bullet)

 (DE)

Namensvorschläge im Deutschen (dabei steht „az" jeweils für „Aufzählungszeichen"):

- **li_liste_intro**
- **l1_liste_ebene1_az**
- **l1f_liste_ebene1_folge**
- **l2_liste_ebene2_az**
- **l2f_liste_ebene2_folge**

This paragraph introduces the list:] ← li_list_intro

- This is the first list item.   ] ← l1_list_level1_bullet

- This is the second list item.
  You can use manual line breaks.   ] ← l1_list_level1_bullet

- This is the third list item.   ] ← l1_list_level1_bullet

   This paragraph follows a list 1 item.   ] ← l1p_list_level1_plain

- This is the fourth list item.   ] ← l1_list_level1_bullet

   – This is a level 2 item.   ] ← l2_list_level2_bullet

   – One more level 2 item.   ] ← l2_list_level2_bullet

      This paragraph follows a list 2 item.   ] ← l2p_list_level2_plain

- This is the fifth list item.   ] ← l1_list_level1_bullet

Bullet shapes

Common bullet types are:

- Square bullet
- Round bullet
- Dash
- Triangle
- Arrow
- Check box

For normal lists, square bullets are usually the best option. If you want to make the bullets less prominent, make them gray. Check boxes are particularly adequate for all sorts of to-do lists and checklists.

Tip:
For level 1, choose a bullet that has a stronger visual weight than the bullet that you choose for level 2. For example, use a square bullet for level 1 and a dash for level 2. The larger, the darker, and the more colorful a bullet is, the stronger is its visual weight.

> ⓘ **Important:** Don't use hands and pointing fingers as symbols. These symbols may be offensive in some cultures.

Bullet color

If you want to make bullets less prominent, make them gray.

If you want to make bullets more prominent, add color.

Tip:
Use your corporate color or the predominant color of your product logo.

This paragraph introduces the list:

- This is the first list item.

- This is the second list item.
  You can use manual line breaks within an item.

- This is the third list item.

Indentation

Define a hanging indent.

Tip:
Use the same indent that you also use for procedures (see *Procedure paragraph styles* 201). You can then use level 2 lists also within procedures and you get a more consistent, harmonious layout.

✖ **No:**

✘ No:

✔ Yes:

If space is tight, you can alternatively use a run-in list. A run-in list is less clearly laid out than a list with a hanging indent, but it's still much clearer than standard body text.

If a picture, a table, a warning, or another element relates to a particular list item, indent this element along with the list item. (This is usually only worth the effort if your authoring tool supports this feature without having to define a separate, indented version of each style.)

✖ No:

✔ Yes:

Special settings

li_list_intro:

- Add some extra space before the paragraph so that the space before the list intro is slightly larger than the space before a body paragraph. This sets the list clearly apart from the rest of the text.
- Set to keep with the next paragraph on the same page.

l1_list_level1_bullet:

- Set to keep all paragraph lines together on the same page.

l1p_list_level1_plain:

- Set to keep with the previous paragraph on the same page.
- Set to keep all paragraph lines together on the same page.
- Add less space before this paragraph than before *l1_list_level1_bullet*.

l2_list_level2_bullet:

- Set to keep with the previous paragraph on the same page.
- Set to keep all paragraph lines together on the same page.

l2p_list_level2_plain:

- Set to keep with the previous paragraph on the same page.
- Set to keep all paragraph lines together on the same page.
- Add less space before this paragraph than before *l2_list_level2_bullet*.

Related rules

Procedure paragraph styles 20↑

2.7.6 Note and warning paragraph styles

Tips provide optional information that helps users to perform a task more efficiently or more comfortably.

Standard notes provide supplemental information and comments of any kind other than tips.

Important notes call the readers' attention to typical obstacles that may prevent them from obtaining the desired results.

Warnings alert readers to potential hazards. Depending on the severity of the hazard, there are several types of warnings: "Caution," "Warning," and "Danger."

Make important notes and warnings particularly eye-catching.

Needed styles:

- **at_tip**
  tip
- **an_note**
  standard note
- **ai_important**
  important note
- **ac_caution**
  warning of the type "Caution"
- **aw_warning**
  warning of the type "Warning"
- **ad_danger**
  warning of the type "Danger"

 (DE)

Namensvorschläge im Deutschen: **at_tipp, ah_hinweis, aw_wichtig, av_vorsicht, aw_warnung, ag_gefahr**

Design tips

Choose one basic design for all notes and warning types, and then vary the visual weight according to the importance and severity of the note or warning:

- From least important to most important, the sequence is:
  tip > note > important note > caution > warning > danger

- You can vary color and borders.
- You can use symbols either with all notes and warnings or only with the more important ones.

If you want to set up a particular design but can't accomplish it with paragraph properties alone, use tables. Remember that you can also use borderless tables.

If you add borders around notes or warnings:

- A good value for the line width is the width of the line in the letter "I" written in the used font and font size. If the line isn't black but gray or colored, increase the line width up to a maximum of the double value.
- A good value for the distance between the border and the text is the width of the lowercase letter "m" in the used font. The bottom margin often needs to be slightly larger so that all margins visually appear to be identical (see *Trust your visual judgment* 35).

If you use color, make sure that everything remains clearly readable even when printed on a black and white, standard office printer. If you use a dark background color, use white for the font color.

Examples

The following pictures show some examples for your inspiration.

If your layout uses a margin column, you can also move the symbols to the margin column.

Tip:
Lorem ipsum dolor sit amet, consectetur adipiscing elit.
Sed fringilla diam et enim adipiscing ac varius risus varius.

Note:
Lorem ipsum dolor sit amet, consectetur adipiscing elit.
Sed fringilla diam et enim adipiscing ac varius risus varius.

WARNING:
Lorem ipsum dolor sit amet, consectetur adipiscing elit.
Sed fringilla diam et enim adipiscing ac varius risus varius.

Tip:
Lorem ipsum dolor sit amet, consectetur adipiscing elit.
Sed fringilla diam et enim adipiscing ac varius risus varius.

Note:
Lorem ipsum dolor sit amet, consectetur adipiscing elit.
Sed fringilla diam et enim adipiscing ac varius risus varius.

 Important: Lorem ipsum dolor sit amet, consectetur adipiscingv elit. Sed fringilla diam et enim adipiscing ac varius risus varius.

 Caution: Lorem ipsum dolor sit amet, consectetur adipiscingv elit. Sed fringilla diam et enim adipiscing ac varius risus varius.

 Note: Lorem ipsum dolor sit amet, consectetur adipiscing elit. Sed fringilla diam et enim adipiscing ac varius risus varius.

Note
Lorem ipsum dolor sit amet, consectetur adipiscing elit.
Sed fringilla diam et enim adipiscing ac varius risus varius.

Note: Lorem ipsum dolor sit amet, consectetur adipiscing elit. Sed fringilla diam et enim adipiscing ac varius risus varius.

CAUTION

Moving parts can snag and pull.
May cause injury.

▶ Do not wear loose clothing or jewelry and pull back long hair in a hairnet or ponytail while operating the machine.

▶ Do not remove the guard.

⚠ WARNING

Moving parts can snag and pull.
May cause severe injury.

▶ Do not wear loose clothing or jewelry and pull back long hair in a hairnet or ponytail while operating the machine.

▶ Do not remove the guard.

⚠ DANGER

Moving parts can snag and pull.
May cause severe injury or death.

▶ Do not wear loose clothing or jewelry and pull back long hair in a hairnet or ponytail while operating the machine.

▶ Do not remove the guard.

(DE)

Im Deutschen lauten die entsprechenden Signalwörter:

Englisch	Deutsch
Tip	**Tipp**
Note	**Hinweis**
Important	**Wichtig**
CAUTION	**Vorsicht!**
WARNING	**Warnung!**
DANGER	**Gefahr!**

Hinweis:
Eine Schreibung in Großbuchstaben ist im Deutschen eher ungebräuchlich, stattdessen stehen häufig Ausrufezeichen.

Special settings

at_tip, **an_note**, **ai_important**, **ac_caution**, **aw_warning**, **ad_danger**:

Set to keep all paragraph lines together on the same page.

2.8 Recommended character styles

Character styles format single words or phrases. They determine the used typeface and color.

Professional authoring tools also allow you to control automatic hyphenation and line breaks in the word or phrase to which a particular character style has been applied.

General recommendations

When setting up character styles:

- Create as few character styles as possible. Users aren't able to distinguish among more than 3 or 4 distinct visual styles (see also *Use clear and simple design* 19).

- Create styles only for tagging particular types of information (see *Create styles semantically* 47).

- When possible, define styles that adhere to general conventions so that readers don't have to learn a new meaning of a known style. For example, it's most common to use:

  - bold font style for controls

  - underlined text for hyperlinks

  - italic font style for special terms and for emphasis

When applying character styles:

- Don't include surrounding spaces and punctuation.

- Don't use character styles in headings and subheadings.

Overview of recommended styles

The basic character styles needed for every user assistance document are:

Purpose	Suggested style name	Suggested keyboard shortcut
To highlight the name of the documented product and to disable automatic hyphenation for the product name (optional). For details, see: *Product name character style* 223	cp_**p**roduct	[Ctrl]+[Shift]+[**p**]

To highlight all user interface and interaction elements, such as window titles, menu items, buttons, keys, levers, and so on. For details, see: *Element character style* 225	ce_element	[Ctrl]+[Shift]+[e]
To mark parameters, such as parameters of function calls, formulas, and so on. For details, see: *Parameter character style* 226	ca_parameter	[Ctrl]+[Shift]+[a]
To mark product-specific technical terms. For details, see: *Term character style* 228	ct_term	[Ctrl]+[Shift]+[t]
To emphasize words and expressions if this is necessary to avoid confusion. For details, see: *Emphasis character style* 230	cm_emphasis	[Ctrl]+[Shift]+[m]
To highlight important keywords that help readers to skim the text (optional). For details, see: *Strong character style* 232	cs_strong	[Ctrl]+[Shift]+[s]
To mark input that users must type. For details, see: *User input character style* 234	ci_input	[Ctrl]+[Shift]+[i]
To mark quotes of source code and configuration files. For details, see: *Code character style* 235	cc_code	[Ctrl]+[Shift]+[c]
To mark links / cross-references to other topics within the document as well as links to external web sites and documents. For details, see: *Link character styles* 237	cl_link	[Ctrl]+[Shift]+[l]

(DE)

Wenn Sie lieber mit deutschen Bezeichnungen arbeiten, können Sie z. B. die folgenden Formatnamen und Tastenkürzel verwenden:

Zweck	Vorschlag für Formatnamen	Vorschlag für Tastenkürzel
Um den Namen des beschriebenen Produkts hervorzuheben sowie um die automatische Silbentrennung für den Produktnamen zu deaktivieren (optional) Details siehe: *Product name character style* 223	zp_**p**rodukt	[Ctrl]+[Shift]+[**p**]
Um alle Schnittstellen- und Bedienelemente zu kennzeichnen, wie Fenstertitel, Menüpunkte, Schaltflächen, Knöpfe, Bedienhebel, usw. Details siehe: *Element character style* 225	ze_**e**lement	[Ctrl]+[Shift]+[**e**]
Um Parameter zu kennzeichnen, wie z. B. Parameter für Funktionsaufrufe, Formeln, usw. Details siehe: *Parameter character style* 226	za_**p**arameter	[Ctrl]+[Shift]+[**a**]
Um produktspezifische Fachbegriffe zu kennzeichnen Details siehe: *Term character style* 228	zi_begr**i**ff	[Ctrl]+[Shift]+[**i**]
Um Wörter und Ausdrücke hervorzuheben, wenn diese Hervorhebung Missverständnissen vorbeugen kann Details siehe: *Emphasis character style* 230	zb_**b**etonung	[Ctrl]+[Shift]+[**b**]
Um wichtige Schlüsselwörter hervorzuheben, die den Lesern helfen, den Text schnell zu	zh_**h**ervorhebung	[Ctrl]+[Shift]+[**h**]

überfliegen (optional) Details siehe: *Strong character style* 232		
Zur Kennzeichnung von Benutzereingaben über die Tastatur Details siehe: *User input character style* 234	zg_ein**g**abe	[Ctrl]+[Shift]+[**g**]
Zur Kennzeichnung von Quellcode-Zitaten und Zitaten aus Konfigurationsdateien Details siehe: *Code character style* 235	zc_**c**ode	[Ctrl]+[Shift]+[**c**]
Zur Kennzeichnung von Links / Querverweisen zu anderen Themen innerhalb desselben Dokuments, zu externen Webseiten oder externen Dokumenten Details siehe: *Link character styles* 237	zl_**l**ink	[Ctrl]+[Shift]+[**l**]

Related rules

Automate line breaks and page breaks 41

Create styles semantically 47

Create styles hierarchically 56

Use color with care 21

Think ahead about printing 36

Choosing fonts and spacing 85

2.8.1   Product name character style

This style can be used to set off the name of the documented product. Having this style is optional. It's purely a matter of design and corporate identity. You can just as well write your product's name as regular body text without any special markup.

Don't use quotation marks around product names, regardless of whether you use any special character style.

Tip:
You can also use a special character style for the product name to suppress hyphenation and line breaks within your product name—even if you don't set up any special visual characteristics.

Suggested style name: **cp_product**

 (DE)

Namensvorschlag im Deutschen: **zp_produkt**

With DemoSoft you can print reports.

With DemoSoft you can print reports.

With DemoSoft you can print reports.

With *Demo Soft* you can print reports.

Recommended design

- If your product uses special colors or fonts, you can also use these colors and fonts for the product name.
- If you don't have any given design that you need to employ, don't invent one. Instead, format your product's name similar to standard body text.

Special settings

- Disable hyphenation.
- Disable line breaks.

Related rules

Element character style 225

Term character style 226

2.8.2   Element character style

This style is needed to mark all user interface and interaction elements, such as window titles, menu items, buttons, keys, switches, levers, and so on.

Suggested style name: **ce_element**

 (DE)

Namensvorschlag im Deutschen: **ze_element**

In the **Font style** list, select **Bold**.

Click the Options button.

From the File menu, choose Print.

Recommended design

- Use a bold version of the standard body font.
- If bold text appears too emphatic, don't make it black but use some shade of gray. This clearly sets it apart and yet keeps it unobtrusive.
- If you want to use color, preferably choose either your corporate color or the predominant color of your product logo.

Special settings

- Disable hyphenation.
- Disable line breaks.

Related rules

Product name character style 223
Term character style 228

2.8.3 Parameter character style

This style is needed to mark parameters, such as parameters of function calls, formulas, and so on.

Suggested style name: **ca_parameter**

 (DE)

Namensvorschlag im Deutschen: **za_parameter**

When you set the temperature to *cool*, energy consumption is slightly higher.

Possible command line parameters are:
`/add` **and** `/replace`.

Possible command line parameters are:
`/add` **and** `/replace`.

Recommended design

- Preferably, use an italic form of the standard body font.
- You can also use a monospaced font (such as Courier) if parameters can't be selected from a given set but must be typed.
- If parameters are important landmarks for which readers will skim your document, make them bold or add some color.

Special settings

- Disable hyphenation.
- Disable line breaks.

▶ **Related rules**

Term character style 228
User input character style 234

Code character style 235

2.8.4 Term character style

This style is optional. It can be used to mark product-specific technical terms. Often, these terms follow words such as "called," "known as," "labeled," "stands for," and so on.

Suggested style name: **ct_term**

 (DE)

Namensvorschlag im Deutschen: **zi_begriff**

Several fonts that share the same typeface are called a *font family*.

The patented ***Marathon Print*** function saves up to 50% toner.

Recommended design

- Use an italic form of the standard body font.
- Don't use quotation marks instead of setting up this character style, and don't use quotation marks around this style. Quotation marks make your text longer than necessary and add clutter that you can avoid.
- If product-specific technical terms are important landmarks for which readers will skim your document, make them bold+italic.

Special settings

- Disable hyphenation.
- Disable line breaks.

Related rules

Product name character style 223
Element character style 225
Parameter character style 226

2.8.5   Emphasis character style

This style is sometimes needed to emphasize words and expressions to avoid misinterpretation or confusion.

> **Important:** Don't use this style to highlight important information. For this purpose, use the *Strong character style* 232.

When writing texts, apply the emphasis character style sparingly to avoid dulling the impact on the reader. Only use the emphasis style when it really makes a sentence clearer. As a decision aid, read the sentence aloud: Use the emphasis style only if you have actually emphasized the word.

Suggested style name: **cm_emphasis**

 (DE)

Namensvorschlag im Deutschen: **zb_betonung**

This character style is *not* dispensable.

Move the shutter to the *left*.

Recommended design

- Use an italic form of the standard body font.
- Don't make the emphasis bold and don't use color. It's sufficient if readers notice the emphasis while reading. The style shouldn't attract attention on its own because it doesn't mark any information that needs to be accessed in a targeted way.
- Don't use ALL CAPS. This style was common for emphases on mechanical typewriters, which didn't have any italic font style. However, like bold or colored text, ALL CAPS is actually too strong for a simple emphasis. In addition, ALL CAPS slows readers down because the words don't have their characteristic shapes.

Special settings

Related rules

Product name character style 223

Element character style 225

Parameter character style 226

Term character style 228

Strong character style 232

2.8.6 Strong character style

This style can be used to highlight important keywords that help readers skim a document for specific information.

> ⓘ **Important:** Don't use this style if you want to emphasize a word to avoid misinterpretation or confusion. In this case, use *Emphasis character style* 230.

When writing texts, apply the strong character style sparingly. If you use it too often, it loses its power.

Suggested style name: **cs_strong**

> ▬ (DE)
>
> Namensvorschlag im Deutschen: **zh_hervorhebung**

If you want to **write a letter**, lorem ipsum dolor sit amet, consectetur adipiscing elit. Morbi erat nunc, auctor eget condimentum eget, viverra a ante.

If you want to **write a report**, praesent sollicitudin, nisl vel elementum pellentesque, dolor libero elementum nunc, sit amet auctor massa turpis ut ipsum. Praesent blandit vestibulum nisl mollis gravida.

Recommended design

- Use a bold form of the standard body font.
- Don't use ALL CAPS or Small Caps. These styles slow readers down because the words don't have their characteristic shapes.

Special settings

Related rules

Product name character style `223`

Element character style `225`

Parameter character style `226`

Term character style `228`

Emphasis character style `230`

2.8.7   User input character style

This style is needed to mark input that users must type.

Suggested style name: **ci_input**

> ▬▬ (DE)
>
> Namensvorschlag im Deutschen: **zg_eingabe**

Into the User name field, type: admin

Into the Password field, type: demo

Recommended design

- Choose a typewriter-like, monospaced font, such as Courier.
- If you want to use color, use a shade of blue that's similar to the blue ink of a pen.

Special settings

- Disable hyphenation.

> **Related rules**
>
> *Parameter character style* 226
> *Term character style* 228
> *Code character style* 235

2.8.8 Code character style

This style is needed to mark:

- quotes from program source code
- examples of program source code
- computer-language elements such as reserved words, variables, procedures, and so on
- commands in configuration files

If your help authoring tool supports automatic syntax highlighting for program source code, use this feature instead of the code character style or in addition to the code character style.

Suggested style name: **cc_code**

 (DE)

Namensvorschlag im Deutschen: **zc_code**

This is an HTML snippet: `<p>Hello world.</p>`

Recommended design

- Use a monospaced font, such as Courier.
- Don't use color. (This doesn't apply if you use automatic syntax highlighting.)
- Don't use any other special formatting.

Special settings

- Disable hyphenation.
- Disable line breaks.

Related rules

Parameter character style 226

Term character style 228
User input character style 234

2.8.9 Link character styles

This style is needed to mark links and cross-references to other topics *within* the document, as well as to mark links to *external* documents and web sites.

Don't use quotation marks instead of this character style because this would make your text longer and less clearly laid out.

Suggested style name: **cl_link**

 (DE)

Namensvorschlag im Deutschen: **zl_link**

In online help:

For detailed information, see Working with Dummies.

In printed documents:

For detailed information, see *"Working with Dummies" on page 26.*

For detailed information, see *Working with Dummies* on page 26.

For detailed information, see Working with Dummies on page 26.

Recommended design

- In online help, preferably make hyperlinks underlined. Because this is the most common style for links, readers can immediately identify underlined text as a link. If you have very good reasons not to underline links, at least design the style so that the text gets underlined when the mouse hovers over the link.

- In printed documents, preferably make cross-references italic. *Don't* underline them here because this would put too much visual weight on them. To distinguish cross-references from other italic text elements (such as parameters, terms, and emphases) a good solution is to use a shade of gray.

  If it's possible with your authoring tool: *Don't* format the page number with the link character style. You then don't need quotation marks around the link text, which removes some unnecessary clutter (compare the examples above).

- If you create printed manuals and online help from the same text base (single source publishing), make sure that the conversion process maps the source format to both output formats correctly. In particular, make sure that page numbers and phrases like "on page" don't appear in online help.

Special settings

Unless you have a very narrow text column, consider disabling hyphenation and disabling line breaks.

Links and cross-references with line breaks are hard to read. Some readers might even interpret a link that's interrupted by a line break as two different links.

3   References

This is the end of the book—but it's not the end of the story.

We hope that this book made you aware of the key factors that account for good user assistance, and we hope that the book will be your guide when you create your own documents.

If you want to learn more, take a look at the other books in our Technical Documentation Solutions Series:

- Technical Documentation Solutions Series: **Planning and Structuring User Assistance** — How to organize user manuals, online help systems, and other forms of user assistance in a user-friendly, easily accessible way

- Technical Documentation Solutions Series: **Writing Plain Instructions** — How to write user manuals, online help, and other forms of user assistance that every user understands

- Technical Documentation Solutions Series: **Illustrating and Animating Help and Manuals** — How to create pictures, instruction videos, and screencasts that communicate technical information clearly

In addition, the following bibliography may be helpful. It also contains the books mentioned above. Flag symbols indicate the language of each book.

Books on technical documentation in general

Achtelig, Marc
Planning and Structuring User Assistance: How to organize user manuals, online help systems, and other forms of user assistance in a user-friendly, easily accessible way. indoition, 2012.

Achtelig, Marc
*Technical Documentation Essentials: "How to Write That F***ing Manual": The essentials of technical writing in a nutshell.* indoition, 2012.

Achtelig, Marc
*Technische Dokumentation: „How to Write That F***ing Manual": Ohne Umschweife zu benutzerfreundlichen Handbüchern und Hilfen.* Zweisprachige Ausgabe Englisch + Deutsch. indoition, 2012.

Achtelig, Marc
Translating Technical Documentation Without Losing Quality: What you shouldn't spoil when translating user manuals and online help. indoition, 2012.

Ament, Kurt
Indexing: A Nuts-and-Bolts Guide for Technical Writers. William Andrew, 2007.

Ament, Kurt
Single Sourcing: Building Modular Documentation. William Andrew, 2002.

Ballstaedt, Steffen-Peter
Wissensvermittlung. Die Gestaltung von Lernmaterial. Beltz Psychologische
Verlags Union PVU, 1997.

Barker, Thomas T.
Writing Software Documentation: A Task-Oriented Approach. Longman, 2002.

Baumert, Andreas
*Interviews in der Recherche: Redaktionelle Gespräche zur
Informationsbeschaffung.* VS Verlag für Sozialwissenschaften, 2004.

Bellamy, Laura; Carey, Michelle; Schlotfeldt, Jenifer
DITA Best Practices: A Roadmap for Writing, Editing, and Architecting in DITA.
IBM, 2011.

Bellem, Birgit; Dreikorn, Johannes; Drewer, Petra; Fleury, Isabelle;
Haldimann, Ralf; Jung, Martin; Keul, Udo P.; Klemm, Viktoria; Lobach, Sabine;
Prusseit, Ines; Reuther, Ursula; Schmeling, Roland; Schmitz, Klaus-Dirk;
Sütterlin, Volker
*Leitlinie Regelbasiertes Schreiben – Deutsch für die Technische
Kommunikation.* tekom, 2010.

Bremer, Michael
*The User Manual Manual: How to Research, Write, Test, Edit & Produce a
Software Manual.* Untechnical, 1999.

Brändle, Max (Hrsg.); Gabriel, Carl-Heinz (Hrsg.); Pforr, Reinhard (Hrsg);
Pichler, Wolfram (Hrsg.); Schmidt, Curt (Hrsg.); Schulz, Matthias (Hrsg.)
Leitfaden für Betriebsanleitungen. tekom, 2010.

Carroll, John M. (Editor)
Minimalism Beyond the Nurnberg Funnel. The MIT Press, 1998.

Carroll, John M.
*The Nurnberg Funnel: Designing Minimalist Instruction for Practical Computer
Skill.* The MIT Press, 1990.

Clements, Paul; Bachmann, Felix; Bass, Len; Garlan, David; Ivers, James;
Little, Reed; Merson, Paulo; Nord, Robert; Stafford, Judith
Documenting Software Architectures: Views and Beyond. Addison-Wesley,
2010.

Clark, Ruth C.
Developing Technical Training: A Structured Approach for Developing Classroom and Computer-based Instructional Materials. Pfeiffer, 2007.

Clark, Ruth C.; Mayer, Richard E.
E-Learning and the Science of Instruction: Proven Guidelines for Consumers and Designers of Multimedia Learning. Pfeiffer, 2011.

Closs, Sissi
Single Source Publishing: Topicorientierte Strukturierung und DITA. entwickler press, 2006.

Coe, Marlana
Human Factors for Technical Communicators. Wiley, 1996.

Cowan, Charles
XML in Technical Communication. Institute of Scientific and Technical Communication, 2010.

DIN e.V (Herausgeber)
Technische Dokumentation: Normen für Produktdokumentation und Dokumentenmanagement. Beuth, 2008.

Drewer Petra; Ziegler, Wolfgang
Technische Dokumentation. Vogel Business Media, 2011.

Ferlein, Jörg; Hartge, Nicole
Technische Dokumentation für internationale Märkte: Haftungsrechtliche Grundlagen, Sprache, Gestaltung, Redaktion und Übersetzung. Expert, 2008.

Garrand, Timothy
Writing for Multimedia and the Web: A Practical Guide to Content Development for Interactive Media. Focal, 2006.

Gentle, Anne
Conversation and Community: The Social Web for Documentation. XML Press, 2009.

Grünwied, Gertrud
Software-Dokumentation: Grundlagen – Praxis – Lösungen. Expert, 2006.

Grupp, Josef
Handbuch Technische Dokumentation: Produktinformationen rechtskonform aufbereiten, wirtschaftlich erstellen, verständlich kommunizieren. Hanser, 2008.

▓ Hackos, JoAnn T.
Information Development: Managing Your Documentation Projects, Portfolio, and People. Wiley, 2007.

▓ Hackos, JoAnn T.
Introduction to DITA: A User Guide to the Darwin Information Typing Architecture. Comtech Services, 2011.

▓ Hahn, Hans-Peter
Technische Dokumentation leichtgemacht. Hanser, 1996.

▓ Hamilton, Richard L.
Managing Writers: A Real-World Guide to Managing Technical Documentation. XML Press, 2008.

▓ Hargis, Gretchen; Carey, Michelle; Hernandez, Ann Kilty; Hughes, Polly; Longo, Deirdre; Rouiller, Shannon; Wilde, Elizabeth
Developing Quality Technical Information: A Handbook for Writers and Editors. IBM, 2004.

▓ Hartman, Peter J.
Starting a Documentation Group: A Hands-On Guide. Clear Point Consultants, 1999.

▓ Hennig, Jörg (Hrsg.), Tjarks-Sobhani, Marita (Hrsg.)
Arbeits- und Gestaltungsempfehlungen für Technische Dokumentation: Eine kritische Bestandsaufnahme. Schmidt-Römhild, 2008.

▓ Hennig, Jörg (Hrsg.); Tjarks-Sobhani, Marita (Hrsg.)
Multimediale Technische Dokumentation. Schmidt-Römhild, 2010.

▓ Hentrich, Johannes
DITA: Der neue Standard für Technische Dokumentation. XLcontent, 2008.

▓ Hoffmann, Walter; Hölscher, Brigitte G.; Thiele, Ulrich
Handbuch für technische Autoren und Redakteure: Produktinformation und Dokumentation im Multimedia-Zeitalter. Publicis, 2002.

▓ Hörmann, Hans
Meinen und Verstehen: Grundzüge einer psychologischen Semantik. Suhrkamp, 1978.

▓ Horn, Robert E.
Mapping Hypertext: The Analysis, Organization, and Display of Knowledge for the Next Generation of On-Line Text and Graphics. Lexington, 1990.

Horton, William
Designing and Writing Online Documentation: Hypermedia for Self-Supporting Products. Wiley, 1994.

Johnston, Carol; Critcher, Ginny; Pratt, Ellis
How to write instructions. Cherryleaf, 2011.

Juhl, Dietrich
Technische Dokumentation: Anleitungen und Beispiele. Springer, 2005.

Kothes, Lars
Grundlagen der Technischen Dokumentation: Anleitungen verständlich und normgerecht erstellen. Springer, 2010.

Kühn, Cornelia
Handlungsorientierte Gestaltung von Bedienungsanleitungen. Schmidt-Römhild, 2004.

Muthig, Jürgen (Hrsg.)
Standardisierungsmethoden für die Technische Dokumentation. Schmidt-Römhild, 2008.

Pearsall, Thomas E.; Cook, Kelli Cargile
Elements of Technical Writing. Longman, 2009.

Price, Jonathan; Korman, Henry
How to Communicate Technical Information: A Handbook of Software and Hardware Documentation. Addison-Wesley Professional, 1993.

Pringle, Alan S.; O'Keefe, Sarah S.
Technical Writing 101: A Real-World Guide to Planning and Writing Technical Documentation. Scriptorium, 2009.

Rockley, Ann; Manning, Steve; Coopern Charles
Dita 101. lulu, 2009.

Rockley, Ann; Cooper, Charles
Managing Enterprise Content: A Unified Content Strategy. New Riders, 2012.

Schriver, Karen A.
Dynamics in Document Design: Creating Text for Readers. Wiley, 1996.

Schwarzman, Steven
Technical Writing Management: A Practical Guide. CreateSpace, 2011.

Self, Tony
The DITA Style Guide: Best Practices for Authors. Scriptorium, 2011.

tekom (Hrsg.)
Richtlinie zur Erstellung von Sicherheitshinweisen in Betriebsanleitungen.
tekom, 2005.

Thiele, Ulrich
Technische Dokumentationen professionell erstellen. WEKA, 2009.

Thiemann, Petra; Krings, David
Creating User-Friendly Online Help: Basics and Implementation with MadCap Flare. CreateSpace, 2009.

Tuffley, Dr. David
Software User Documentation: A How To Guide for Project Staff. CreateSpace, 2011.

Van Laan, Krista; Julian, Catherine; Hackos, JoAnn
The Complete Idiot's Guide to Technical Writing. Alpha, 2001.

Weber, Jean Hollis
Is the Help Helpful? How to Create Online Help That Meets Your Users' Needs. Hentzenwerke, 2004.

Weber, Klaus H.
Dokumentation verfahrenstechnischer Anlagen: Praxishandbuch mit Checklisten und Beispielen. Springer, 2008

Weiß, Cornelia
Professionell dokumentieren. Beltz, 2000.

Weiss, Edmond H.
How To Write Usable User Documentation. Oryx, 1991.

Welinske, Joe
Developing User Assistance for Mobile Apps. Lulu, 2011.

Wieringa, Douglas; Barnes, Valerie E.; Moore, Christopher
Procedure Writing: Principles and Practices. Battelle, 1998.

Young, Indi
Mental Models: Aligning Design Strategy with Human Behavior. Rosenfeld, 2008.

Books on graphics and design

Achtelig, Marc
Designing Templates and Formatting Documents: How to make user manuals and online help systems visually appealing and easy to read, and how to make templates efficient to use. indoition, 2012.

Alexander, Kerstin
Kompendium der visuellen Information und Kommunikation. Springer, 2007.

Ballstaedt, Steffen-Peter
Visualisieren: Über den richtigen Einsatz von Bildern. UTB, 2011.

Clark, Ruth C.; Lyons, Chopeta
Graphics for Learning: Proven Guidelines for Planning, Designing, and Evaluating Visuals in Training Materials. Pfeiffer, 2010.

Cooper, Alan
About Face: The Essentials of User Interface Design. IDG, 1999.

Cooper, Alan
The Inmates Are Running the Asylum. SAMS, 1999.

Hennig, Jörg (Herausgeber); Marita Tjarks-Sobhani (Herausgeber)
Visualisierung in Technischer Dokumentation. Schmidt-Römhild, 2004.

Horton, William
Illustrating Computer Documentation: The Art of Presenting Information Graphically on Paper and Online. Wiley, 1991.

Runk, Claudia
Grundkurs Grafik und Gestaltung. Galileo, 2010.

Williams, Robin
The Non-Designer's Type Book. Peachpit, 2005.

Williams, Robin
The Non-Designer's Design Book. Peachpit, 2008.

Wirth, Thomas
Missing Links: Über gutes Webdesign. Hanser, 2004.

4 Feedback

We sincerely hope that reading this book was a rewarding experience.

- If you like this book and think that it can help you improve your own documents, please don't hesitate to post a review and recommend the book to your colleagues. Also, don't hesitate to drop us a line. It motivates us so much to carry on :-).

- If you didn't like this book—we're embarrassed and awfully sorry. Could you please send us some feedback about what you think we should improve?

Our email address is: *feedback-DES-DE-1@indoition.com*

If you'd like to help us even more, please also email us your answers to the questions below.

Thank you for your support.

How to answer the questions

Please email your answers to:
feedback-DES-DE-1@indoition.com

For example, your email could look like this:
1c, 2a, 3b, 4a, 5d, 6a, 7a, 8b, 9b, 10a, 11a, 12b

We won't use your data for any purpose other than improving future editions of this book. If you don't want to answer all questions, that's perfectly OK. Just answer the ones that you feel comfortable with.

1. Questions about the book

How did you feel about the length of the book?

It was much too long.	1a
It was slightly too long.	1b
It was just perfect.	1c
It was slightly too short.	1d
It was much too short.	1e

Did the book cover what you'd expected, based on its title and description?

I didn't miss anything.	2a
I missed a few minor things.	2b
I missed some important points.	2c

How did you experience the depth of information?

Much of the presented information was too trivial for me.	3a
The information was just what I needed.	3b
Much of the information was too specialized for me.	3c

How did you like the practical nature of the book?

I appreciated the lack of theory and technical terms.	4a
I missed scientific background information, references to studies, and more precise terminology.	4b

Did you find any mistakes?

Yes, too many.	5a
Some, but not more than usual.	5b
Only very few.	5c
None.	5d

(If your answer is "None": Go through the book again before answering this question! No book is free from errors. If you have the time, please tell us more about the mistakes that you've found.)

2. Questions about your professional background

What's your main professional occupation?

technical writing	6a
support	6b
development	6c
marketing	6d
product management	6e
translation	6f
other	6g

How many years of experience in technical writing do you have?

less than 1	7a
1 to 3	7b
more than 3	7c

Which kind of products do you document?

mainly hardware	8a
mainly software	8b
a mixture of both hardware and software	8c

Who reads the documents that you write?

mainly consumers	9a
mainly professional users	9b

Do you speak English as a first language?

English is my first language.	10a
I speak English as a second language.	10b

Do you mainly write in English?

Yes, more than 50% of my texts are in English.	11a
No, less than 50% of my texts are in English.	11b
No, I don't write English documents at all.	11c

Who purchased this book?

I purchased the book at my own expense.	12a
The organization that I work for purchased the book.	12b
I borrowed the book from a colleague.	12c
I borrowed the book from a public library.	12d
I received a copy in a training course.	12e

Index

Weitere Bücher von indoition publishing zum Thema Technische Dokumentation:

„How to Write That F*ing Manual"**

Ohne Umschweife zu benutzerfreundlichen Handbüchern und Hilfen

Zweisprachige Ausgabe: Englisch + Deutsch

„Writing Plain Instructions"

Wie Sie Handbücher, Online-Hilfen und andere Formen Technischer Kommunikation schreiben, die jeder Benutzer versteht

Zweisprachige Ausgabe: Englisch + Deutsch

„Planning and Structuring User Assistance"

Wie Sie Handbücher, Online-Hilfen und andere Formen Technischer Dokumentation benutzerfreundlich aufbauen und den Informationszugriff erleichtern

Zweisprachige Ausgabe: Englisch + Deutsch

„Illustrating and Animating Help and Manuals"

Wie Sie Bilder, Instruktionsvideos und Screencasts erstellen, die technische Informationen verständlich vermitteln

Zweisprachige Ausgabe: Englisch + Deutsch

„Dokumentation verlustfrei übersetzen"

Was Sie beim Übersetzen von Benutzerhandbüchern und Online-Hilfen nicht zerstören sollten

Zweisprachige Ausgabe: Englisch + Deutsch

Detaillierte Informationen zu allen Ausgaben finden Sie unter *www.indoition.de.*

Copy and Paste Kit Technische Dokumentation
Ihre Bausteine für verständliche Dokumentation

Das Technical Documentation Copy and Paste Kit ist eine wesentlich erweiterte Online-Version dieses Buchs sowie aller weiterer Bücher der Reihe „Lösungen zur Technischen Dokumentation". Das Kit ist Ihr steter Begleiter und Styleguide (Redaktionsleitfaden) während **aller Phasen eines Dokumentations-Projekts**:

- Analyse der Anforderungen
- Strukturierung der Inhalte
- Design von Formatvorlagen
- Schreiben der Texte
- Anfertigen von Bildern
- Lektorat und Korrektur
- Übersetzung

Wie dieses Buch, enthält auch das Kit keine langen theoretischen Abhandlungen. Dafür bietet es praxisnahe Empfehlungen und Beispiele, die Sie einfach kopieren und direkt auf Ihre eigene Arbeit anwenden können.

Sie können das Kit entweder lokal, auf einem Netzlaufwerk oder auf einem Webserver installieren. Somit haben Sie und Ihr Team **jederzeit und von überall aus darauf Zugriff**.

Sie können sogar **Ihre eigenen, firmenspezifischen Anmerkungen und Spezifikationen hinzufügen**. Um Ihre Anmerkungen zu bearbeiten und zu verwalten, können Sie nahezu jeden HTML-Editor, ein Wiki, ein Commenting Script oder ein Content-Management-System verwenden. Sie können entweder allen Mitgliedern Ihres Teams das Bearbeiten von Notizen und Kommentaren erlauben oder hierfür einen Moderator bestimmen. Wenn Sie ein Update des Kits installieren, bleiben Ihre Notizen und Kommentare vollständig erhalten.

Auf diese Weise erhalten Sie de facto Ihren eigenen, **firmenspezifischen Styleguide**, ohne sich in wesentlichen Teilen um dessen Erstellung und Pflege kümmern zu müssen.

Mehr Informationen sowie eine Demo finden Sie unter *www.indoition.de*.

indoition Hotkey Script Collection for Writers and Translators

Zeitsparende Makros zum Schreiben und Nachschlagen in jedem Programm

Die Scripts der „indoition Hotkey Script Collection for Writers and Translators" machen Ihre Arbeit effizienter:

- Geben Sie **häufig verwendete Wörter und Wendungen automatisiert** mit Hilfe bestimmter Tastenkombinationen ein.
- **Tippen Sie Sonderzeichen komfortabel mit einem einzigen Tastendruck**, z. B.: sprachspezifische Sonderzeichen, diakritische Zeichen, typografische Anführungszeichen, typografische Apostrophe und Gedankenstriche.
- **Verwandeln Sie die Feststelltaste** in eine reguläre Umschalttaste, so dass SO ETWAS NIE MEHR PASSIERT, wenn Sie versehentlich die Feststelltaste drücken.
- **Schlagen Sie ein markiertes Wort per Tastendruck** in jedem beliebigen Online-Wörterbuch oder Online-Lexikon **nach**.
- Und vieles mehr ...

Alle Scripts können Sie bei Bedarf einfach bearbeiten und individuell anpassen. Dazu brauchen Sie keine fortgeschrittenen Programmierkenntnisse.

Anders als Makros, die für eine bestimmte Applikation programmiert wurden (wie z. B. Microsoft-Word-Makros), funktionieren die Scripts der Script Collection in *allen* Programmen für Windows.

Mehr Informationen sowie eine Demo finden Sie unter *www.indoition.de*.

indoition Starter Template

Professionelle Formatvorlage für Technische
Dokumentation

Viele Autorenwerkzeuge werden ohne geeignete Formatvorlagen zum
Erstellen klarer, ansprechender Technischer Dokumentation geliefert. Eine
eigene Formatvorlage komplett neu zu erstellen, kann zeitaufwendig sein.
Das „indoition Starter Template" beschleunigt diese Arbeit und hilft Ihnen,
kostspielige strategische Fehler von vornherein zu vermeiden. Es bietet:

- Ein Design, das nicht nur angenehm fürs Auge ist, sondern auch Ihre
  Inhalte klar kommuniziert.

- Absatzformate und Zeichenformate, die beim Schreiben Ihrer
  Dokumente einfach anzuwenden sind.

Das Starter Template wurde für Microsoft Word, OpenOffice und LibreOffice
und die Papierformate A4 und Letter entwickelt. Falls Sie ein anderes
Papierformat nutzen, müssen Sie im Wesentlichen nur die Einstellungen für
die Seitenränder ändern.

Auch viele andere Autorenwerkzeuge können Microsoft-Word-Dateien
(*.docx) und die von OpenOffice / LibreOffice genutzten OpenDocument-
Text-Dateien (*.odt) importieren.

Wesentliche Merkmale:

- **Kein Schnickschnack** – die Formatvorlage enthält nur das, was Sie und
  Ihre Leser wirklich brauchen.

- **Automatisierte Formate** machen manuelle Formatierungen
  weitgehend überflüssig. Optimierte Einstellungen sorgen für
  automatische Zeilenumbrüche und Seitenumbrüche.

- **Bewährtes**, systematisches Schema für Formatnamen und
  Tastenkürzel.

- **Funktioniert mit allen Sprachversionen von Microsoft Word,
  OpenOffice und LibreOffice**. Sie müssen keine Formatnamen und
  Feldfunktionen anpassen.

- **Vorbereitet** auf die Möglichkeit, aus Ihren Dokumenten mit Hilfe eins
  geeigneten Single-Source-Publishing-Tools oder Konverters auch **Online-
  Hilfen** erzeugen zu können.

- Enthält eine **ausführliche Anleitung**, wie Sie die Formate anwenden
  und bei Bedarf ändern können.

Mehr Informationen unter *www.indoition.de*.

This is *not* a happy customer:

Make them happy, write better help!

Help+Manual®

www.helpandmanual.com

Help+Manual creates all standard online help formats including **HTML Help, Webhelp, PDF manuals** and **e-books** from one single source.

And it's as easy to use as a word processor.

Learn more on our website *http://www.helpandmanual.com*!